WOMEN ENTREPRENEURS IN SMALL AND MEDIUM ENTERPRISES

The logo is an original drawing by Armand ABPLANALP (signed A.A.), donated by the artist.
Le logo est un dessin original de Armand ABPLANALP (signé A.A.), offert par l'artiste.

ORGANISATION FOR ECONOMIC CO-OPERATION AND DEVELOPMENT

ORGANISATION FOR ECONOMIC CO-OPERATION AND DEVELOPMENT

Pursuant to Article 1 of the Convention signed in Paris on 14th December 1960, and which came into force on 30th September 1961, the Organisation for Economic Co-operation and Development (OECD) shall promote policies designed:

- to achieve the highest sustainable economic growth and employment and a rising standard of living in Member countries, while maintaining financial stability, and thus to contribute to the development of the world economy;
- to contribute to sound economic expansion in Member as well as non-member countries in the process of economic development; and
- to contribute to the expansion of world trade on a multilateral, non-discriminatory basis in accordance with international obligations.

The original Member countries of the OECD are Austria, Belgium, Canada, Denmark, France, Germany, Greece, Iceland, Ireland, Italy, Luxembourg, the Netherlands, Norway, Portugal, Spain, Sweden, Switzerland, Turkey, the United Kingdom and the United States. The following countries became Members subsequently through accession at the dates indicated hereafter: Japan (28th April 1964), Finland (28th January 1969), Australia (7th June 1971), New Zealand (29th May 1973), Mexico (18th May 1994), the Czech Republic (21st December 1995), Hungary (7th May 1996), Poland (22nd November 1996) and the Republic of Korea (12th December 1996). The Commission of the European Communities takes part in the work of the OECD (Article 13 of the OECD Convention).

FOREWORD

The OECD organised an International Conference on 16-18 April 1997 on "Women Entrepreneurs in SMEs: A Major Force in Innovation and Job Creation", which brought together government leaders, senior policy makers, small-business entrepreneurs, academics and experts from OECD Member and non-member countries. Their objective was to:

◊ increase understanding about the economic and social contributions of women entrepreneurs in OECD, developing and transition economies, and, through dialogue stimulate the exchange of experiences;

◊ discuss the common issues and challenges facing women-owned SMEs, and identify "best practice" policies and institutions around the world from which to draw lessons;

◊ analyse the main problems and develop policy recommendations in five areas that are of critical importance for fostering the growth of women-owned SMEs: access to financing, access to markets, the business environment and technology, globalisation and international trade, and management issues; and

◊ recommend private and public sector initiatives to facilitate the growth and development of women-owned SMEs globally in OECD countries and developing and transition economies.

The conference was structured around six workshops and an opening and closing plenary session that put the fundamental facts about women-owned businesses into context, and presented the workshop findings and conference conclusions. The workshops addressed the following major issues:

⇒ Workshop 1: Issues in the Business Environment and Technology

⇒ Workshop 2: Globalisation and International Trade

⇒ Workshop 3: Management Issues

⇒ Workshop 4: Dialogue among OECD Countries, Transition and Developing Economies: Future Challenges for Policy Makers

⇒ Workshop 5: Financing Issues

⇒ Workshop 6: Access to Markets

The present publication brings together the highlights of the conference. It comprises a synthesis chapter based on the conference presentations and discussions and the set of recommendations which emerged. It also includes some of the key messages delivered by Ministers and senior policy makers and a selection of papers, representative of the various themes covered at the workshops and the plenary sessions. The conference programme appears at the end of the report.

This publication has been prepared at the OECD by the Secretariat to the Industry Committee's Working Party on SMEs of the Directorate for Science, Technology and Industry, with the assistance of the Secretariat to the Development Assistance Committee of the Directorate for Development Co-operation. It is published on the responsibility of the Secretary-General of the OECD.

TABLE OF CONTENTS

Message from Joanna SHELTON, Deputy Secretary-General of the OECD

Half of the brainpower on Earth is in the heads of women. Today, the difficulty is to move from the acceptance of equal rights to the reality of equal opportunity. This transition will not be complete until women and men have equal opportunities to occupy positions in power structures throughout the world. This includes not only public law-making and policy-formulating bodies, but also the world of private business. Society as a whole stands to gain by accelerating the process.

Indeed, evidence gathered in recent years demonstrates that SMEs owned by women are growing at a faster pace than the economy as a whole in several OECD countries. To benefit fully from the entrepreneurial capacities of women, it is important to understand the structures and attitudes in our societies that support or hinder their efforts. It is equally important to understand whether women bring different approaches to the management of entrepreneurial activity which should be encouraged and strengthened.

The international conference on "Women Entrepreneurs in SMEs" hosted in April 1997 by the OECD in partnership with IBM, enabled participants to explore these fundamental issues and further analyse other important questions, such as: What are the reasons behind the surge in women entrepreneurship? What are the lessons to be learned from exchanges of countries' experiences? For business of all kinds? For managers in the public sector? How can our economies fully benefit from the contributions of women-owned enterprises?

The set of recommendations emerging from the conference include actions to be taken by government, business, financial institutions and associations. They also aim to pave the way for further progress, to which the OECD can contribute through studies and advice to Member governments. The OECD is well-placed to further our knowledge about women entrepreneurship with a view to increasing the overall effectiveness of policies. In this context, the OECD should play a leadership role, and in particular encourage and facilitate the collection and standardisation of statistics and data on SMEs, including women-owned SMEs on a worldwide basis. It should also help Member governments by reviewing the major barriers to women-owned businesses in the fields of finance, and regulatory and administrative burdens, and, in addition, by assessing how technology use should be promoted to improve their competitiveness and networking.

A follow-up conference in the year 2000 will allow us to compare how far the recommendations of this conference have been successfully implemented and the progress which still needs to be made as we move into the 21st century.

Joanna R. Shelton

Message from Cherie PIEBES,
IBM Global Marketing Executive, Women Entrepreneurs

IBM is proud to have served as primary sponsor of the OECD's historic international forum on Women Entrepreneurs in Small and Medium Enterprises (SMEs). In partnership with other multinational corporations, we funded major research in the United States conducted by the National Foundation for Women Business Owners. This research shows that US women entrepreneurs are going into entrepreneurship at a rate twice as fast as men, and they may own 40 per cent of all the small businesses in the country by the year 2000. The research also shows there are now 7.95 million women entrepreneurs in the United States who generate US$2.3 trillion in turnover and employ one out of every four people working in private business.

These statistics have given women business owners growing clout with policy makers interested in how women entrepreneurs can generate jobs and contribute to economic growth, and with corporations interested in business-to-business marketing to women entrepreneurs (such as Wells Fargo Bank in California, which has created a US$10 billion loan programme for US women entrepreneurs).

Over the past year, as IBM has become increasingly aware of the global growth of women entrepreneurs, we have established a world-wide marketing organisation focused on women entrepreneurs, building on programmes established in the United States and Canada several years ago. In addition to developing solutions to meet the specific needs of women entrepreneurs and SMEs, we now have senior women executives responsible for developing partnerships with associations of women entrepreneurs in North America, Latin America, Europe, the Middle East, Africa, and the Asia Pacific region.

IBM's "women entrepreneurial champions" look forward to:

- working with policy makers to improve data and statistics about women entrepreneurs;
- working with researchers to pinpoint research gaps about women entrepreneurs;
- partnering with associations on the strategic use of technology to help women entrepreneurs grow their businesses and expand into new markets, including international markets;
- helping to get banks and other corporations interested in furthering the growth of women-owned businesses;
- helping to accelerate the identification and diffusion of "best practices" when working with women entrepreneurs in order to catalyse a great leap forward for the global community.

IBM pledges to play an active role in supporting the efforts of women entrepreneurs world-wide, both now and in the future.

We are confident that OECD Member countries will recognise the significant impact of this market on the world economy, and that they will provide the necessary support to facilitate continued growth.

Onward and upward with women's business enterprise!

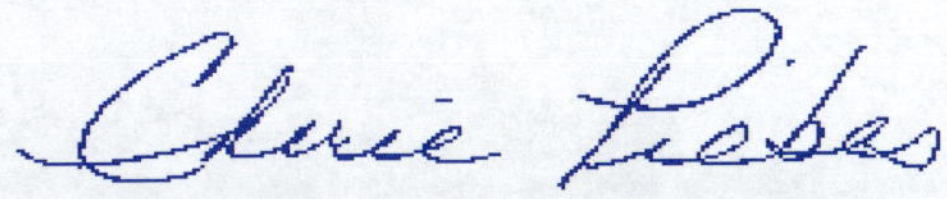

Cherie Piebes

WOMEN ENTREPRENEURS IN SMES: KEY MESSAGES

In organising this conference, the OECD was honoured by the presence of a number of Ministers and senior policy makers. The following quotations, which follow the order of the programme, were chosen to illustrate the importance attached by governments to the issues of women entrepreneurship and SMEs. Extracts from the speeches of the Secretary-General and the Deputy Secretary-General of the OECD are also included, as well as those of a high-level representative of IBM, a major sponsor of the conference. The extracts provide the essential message coming out of the conference, and it is hoped that they will stimulate thinking and provoke constructive change in the field of women entrepreneurs in SMEs.

- ***Mr. Donald J. Johnston, Secretary-General, OECD***

"Half of the brainpower on Earth is in the heads of women. (...) Today, the difficulty is to move from the acceptance of equal rights to the reality of equal opportunity. This transition will not be complete until women and men have equal opportunities for occupying positions in power structures throughout the world. This includes not only public law-making and policy-formulating bodies, but also the world of private business. Society as a whole stands to gain by accelerating the process."

- ***M. René Monory, President of the Senate, France***

"Over the coming decade, the world may well change as much as it did over the last 150 years. This will be due in particular to the immense progress that has been made in communication and technology. We are witnessing a veritable revolution in employment and industrial structures linked to the phenomenon of globalisation. Society is becoming increasingly tertiary, work is becoming more human, and this quite naturally leaves women and SMEs significant space. There should be more French women who leave to work in foreign countries because they have an increasingly important role to play on the international scene."

- ***Ms. Anna Finocchiaro, Minister for Equal Opportunity, Italy***

"The Italian Government has adopted as its strategic point the promotion of new growth through the valorisation of innovating factors represented by women's skills and entrepreneurship. The Government adopted a directive from Prime Minister Romano Prodi on 7 March 1997, announcing 'Actions to promote women's empowerment and to recognise and ensure freedom of choice and a better quality of social living for women and men'. These objectives are in line with the Platform for Actions adopted by the 4th UN World Conference on Women in Beijing (September 1995)"

- ***Mr. Heinrich Kolb, Secretary of State, Ministry of Economy, Germany***

"We need to improve the environment for SMEs, because they are creating scope for more growth employment, broadspace innovation, better skills in the local and global markets. SMEs have to be more forward-looking. In Germany, those women who use government incentives are doing as well as men in creating enterprises. Women bring fresh motivation. Women will fit better into the new service society than in the old industrial society. Our countries can no longer do without the expertise, skills and experience of women entrepreneurs."

- ***Ms. Ann Diamantopoulou, Secretary of State, Ministry of Development, Greece***

"Women and SMEs constitute the main weapons for helping us to build a future without discrimination. We fight for our rights, and not for privileges, because business has no gender. What is a natural success for men is a conquest for women. (...) Women's participation in any kind of economic activity is of a complementary nature to their family incomes; their participation in no way reduces their family duties."

- ***Mme Rita Dionne-Marsolais, Ministry of Industry and Commerce, Quebec, Canada***

"In creating networks of women entrepreneurs we do not seek to marginalise women, but rather to give them access to the new alliances that are shaping tomorrow's economy. Networking gives women the opportunity to end their isolation and, above all, gives them access to strategic information. Beyond any action that we can implement, there is a common denominator that might incite women to create their enterprise: developing an entrepreneurial culture. A taste for business must come from the business environment itself. The OECD can play the role of a facilitator and even of a catalyst in putting these structures into place."

- ***Ms. Betsy Myers, Director of the White House Office of Women's Initiatives and Outreach, United States***

"Women's equality is defined by women's economic empowerment. And the ultimate empowerment is through entrepreneurship. Women-owned businesses are the fastest growing force in the US economy, prompting President Clinton to call women business owners 'the new face of our economy'. Since 1992, our Small Business Administration (SBA) has increased its loans to women by almost 300 per cent. More than 40 per cent of all micro loans are now going to women. (...) When women thrive, their families thrive and the nation thrives."

- ***Mr. Michel Boussard, Vice President, Global General Business Marketing & Business Development, IBM Euro Co-ordination***

"IBM has created a world-wide organisation focused on women entrepreneurs (headed by Cherie Piebes). IBM's four priorities in this field are:

◊ working with policy makers to improve data and statistics on women business owners (WBOs);

◊ partnering with associations of women entrepreneurs on the strategic use of technology to gain competitive advantage and to help WBOs to expand into new markets;

◊ forming global strategic alliances of corporations interested in fostering the growth of WBOs;

◊ accelerating the diffusion and identification of best practices relevant to women."

- ***Ms. Barbara Prammer, Federal Minister for the Condition of Women, Austria***

"This OECD conference clearly signals the public awareness of women's role in the economy and of an emerging shift in the distribution of power. A new generation of highly educated and motivated women is emerging, and they are creating businesses through their own choice and not by family lineage. As newly appointed Minister for the Condition of Women in Austria, I opened the first International Women's Business Centre in Vienna at the beginning of April 1997. The centre's main objective is to minimise the burden that women face in starting a business, by providing professional bankers, trainers and lawyers who can be consulted for advice."

- ***Mme Anne-Marie Couderc, Deputy Minister for Employment in Charge of Women's Rights, France***

"Women provide an essential opportunity for economic and social development and progress. They have a different vision thanks to the delays they are currently remediating, a vision which is new and different, realistic, modern, and enthusiastic. When social structures leave them no possibility for evolving in their careers, and when civil society closes its doors to them, women take their own initiatives. They are quite naturally drawn to initiative, to creation and to the management of businesses.

Economic and social responsibility make political responsibility easier to take. There will be a natural and inevitable progression from one to the other which will be more successful than any discriminatory mechanism; therefore, non-discrimination should remain a basic principle."

- ***The Right Honourable Ms. Irene Pivetti, Former President of the Parliament, Italy***

"This conference reflects ongoing structural change in the economy, in society and in politics. This year we are celebrating the 50th anniversary of the Marshall Plan which gave birth to the OECD. The time is also ripe for change in the OECD, and this conference is an indication of an institution trying to keep pace with social and political realities such as the new role of women as entrepreneurs.

Spreading entrepreneurship in general, and women's entrepreneurship in particular, has always relied on local culture and local social capital. Where entrepreneurship thrives, it is because local values and political powers implement a policy conducive to an enhanced role for women and entrepreneurship in general."

- ***Ms. Joanna R. Shelton, Deputy Secretary-General, OECD***

"A major message that came out across-the-board is the need for government programmes and private sector services to be more efficient in meeting the needs of the users – SMEs in general, and women-owned SMEs in particular.

The recommendations coming out of the Conference are oriented towards the future. They very much centre on how to improve the opportunities available to women to make use of new instruments, exploit and demand new technologies, enter new markets through liberalisation and regulatory reform, do away with costly and ineffective bureaucracy and administrative burdens, and develop a new spirit of entrepreneurship."

Part I:

Synthesis

Chapter 1

WOMEN ENTREPRENEURS IN SMEs

SYNTHESIS

The OECD Secretariat presented an issues paper at the opening of the Conference on "Women Entrepreneurs in SMEs: A Major Force in Innovation and Job Creation", outlining the set of issues that the Conference was expected to address. This Synthesis Chapter is a follow-up to that issues paper and is organised along the same themes. It builds on the wealth of information provided by the Conference in an attempt to present some of the outstanding issues and how they can be resolved. The chapter reflects the facts, ideas and recommendations which emerged from the Conference and highlights the need for comprehensive, internationally comparable quantitative and qualitative data in this area.

Introduction

Women make up half of the world's human resources.

Women make up half of the world's human resources. This seemingly obvious fact has many dimensions, which must be viewed in the current global economic and social context. As industrialised nations struggle to achieve economic growth and job creation, human-resource management, creativity and innovation are the cornerstones of an agenda for improved social and economic conditions.

SMEs are an important source of employment generation, innovation and economic development.

OECD Member and non-member countries are undergoing deep-seated changes and are increasingly challenged by the globalisation of markets, increased competition and the growing interdependence of economies. At the same time, empirical evidence from most OECD countries shows that SMEs are an important source of employment generation, innovation and economic development. This potential tends to remain largely undeveloped, however, and many policy challenges arise because of market imperfections, inappropriate policies and institutional inconsistencies that discriminate against SMEs (OECD, 1996*a*).

Women-owned SMEs are reported to be growing at a faster rate than the economy as a whole in several OECD countries; however, the removal of a number of obstacles would allow their potential to be fully tapped.

This is particularly true for women-owned SMEs, which are reported to be growing at a faster rate than the economy as a whole in several OECD countries. Although women entrepreneurs are rapidly becoming a major force in OECD Member countries, their contribution could become even more significant if a number of restrictions were removed so that their potential could be fully tapped. Their contributions and needs must be examined, and the commensurate structural reforms applied, for three main reasons:

- ◊ ***Economic***: Women-owned SMEs are creating employment for themselves and/or others. Many companies invest in their female staff members over the long term, but do not promote them to the highest management positions – the "glass ceiling" effect. Providing opportunities for these women to branch out and create their own firms is a way of capitalising on their acquired skills and training. Self-employment can also serve as an escape from unemployment.
- ◊ ***Social:*** As well as representing an alternative to unemployment, enterprise creation can also offer women the possibility of balancing work and family responsibilities, thus contributing to family well-being and improving social cohesion. This being said, balancing career and family responsibilities can present a challenge for women entrepreneurs.
- ◊ ***Political:*** Encouraging women-owned SMEs will help reduce the disparities between women and men, increase women's autonomy and allow them to play a more active and representative role in the economic and political life of countries.

Fostering the participation of women entrepreneurs strengthens the economy and can be a source of political, economic and social innovation. Women business owners, as compared to their male counterparts, often have a different vision of organisation and management, because of different experience and outlook. These constitute a real potential source of innovation in terms of management style, company structure, community service, the use of technology, etc. By making recommendations to policy and decision makers, the OECD's objective is to assist Member and non-member countries in realising the innovative and job-creating potential of women-owned SMEs.

"There is a major challenge for policy makers to improve the conditions for the establishment and growth of women-owned SMEs. Governments by themselves cannot make firms grow and employ staff. However, they can work to put in place conditions and incentives which allow firms and individuals to do the job." **[Thomas Andersson, OECD (DSTI), First Plenary Session]**

This synthesis of the Conference on "Women Entrepreneurs in SMEs: A Major Force in Innovation and Job Creation" is organised thematically, as follows:

- ◊ Improving knowledge about female entrepreneurship.
- ◊ Promoting entrepreneurial education and training.
- ◊ Removing financial obstacles and promoting innovative tools to foster the creation and growth of women-owned SMEs.
- ◊ Promoting the competitiveness of women-owned SMEs.
- ◊ Facilitating access to international and domestic markets.
- ◊ Transforming the regulatory and institutional environment.

A. Improving Knowledge about Female Entrepreneurship

Key Issues

One of the major findings was that, in spite of the wealth and variety of information that came out of the Conference, there is clear lack of appropriate and comparable statistics and data about women entrepreneurs.

Data problems

"Women entrepreneurship suffers from a lack of statistical information and research, thus limiting analysis and leading to the use of *a priori* estimates and hypotheses. It is crucially important to promote the development of research in order to better inform the different actors and decision makers about a social and economic phenomenon with a real potential for innovation and job creation in the 21st century." **[Bertrand Duchéneaut, France, First Plenary Session]**

Collecting information and statistics on women business owners should be an integral part of the ongoing data collection on SMEs.

There is general agreement that improved data and statistics on women-owned SMEs are crucial to a better understanding of the contributions of women entrepreneurs to economic growth, international trade and job creation in all countries. Collecting information and statistics on women business owners should be an integral part of the ongoing data collection on SMEs.

Data are often incomplete ...

Indeed, it is difficult to determine the number of women entrepreneurs from official statistics in any country, given often incomplete figures based on sample analyses or surveys. No single formal definition of an SME exists since there is no clear analytical threshold for the concept. Firms can be classified by their turnover, the number of employees and/or their legal status. Statistics regarding women-owned businesses are even more problematic as most surveys ignore gender. Definitional issues further complicate data collection: a "women entrepreneur" can refer equally to someone who has started a one-woman business, to someone who is a principal in a family business or partnership, or to someone who is a shareholder in a publicly held company which she runs.

... and, partly because of historical factors or civil liberties, it is difficult to obtain national statistics on income and wealth by gender.

This situation results partly from historical factors or civil liberties which prohibit national statistical systems from collecting and publishing certain information. Detailed statistics on income and wealth by gender are therefore particularly difficult to obtain from some countries. The organisation of international data collection for purposes of comparison raises further problems of access and coherence.

Raising the visibility of and awareness about the economic and social role of women entrepreneurs and their businesses requires data and statistics.

Again, as mentioned above, quality information and statistics are essential for sound analysis and policy, and for monitoring the effectiveness of initiatives. Raising the visibility of and awareness about the economic and social role of women entrepreneurs and their businesses requires data and statistics for a clearer picture of this phenomenon whose emergence could lead to real possibilities for innovation and jobs in the 21st century. Although data alone are insufficient to establish the credibility of women entrepreneurs, they provide a basis for informed discussion.

"Credibility is the underlying barrier to access to markets, to marketing information, to relationships and to the status of women in the business community. (...) With a group of women business owners, we decided that if we were to be taken seriously, we must have data and statistics that fully represented what we were doing. We established an organisation on our own to develop and highlight the accurate economic contributions of women. In 1992, The National Foundation for Women Business Owners (NFWBO) released its first study, the first accurate picture of women-owned businesses in America, and it has shattered many of the firmly established myths. We know from the data that women are players in economic health and economic stability. As a result of this data, the US Federal Government set a goal of 5 per cent for all federal procurement to go to women-owned businesses. Corporations have actively reached out to involve women as vendors. Access to credit has changed as a result of this data. Wells Fargo has established a US$10 billion-fund for loans to women business owners over a period of ten years. The report has also had a major impact on the access that women have to information, to training and to networks. We can use our data as empowerment to encourage women to start businesses, to change the paradigm in which women operate their businesses." **[Laura Henderson, United States, Access to Markets]**

What do the available data show?

Available data nonetheless point to the growing importance of women-owned enterprises in OECD Member countries.

Participation of women in self-employment gives one indication of countries' female entrepreneurship (see Annex 1). This share has changed in contrasting ways in a number of OECD countries over the past 20 years. During the last ten years (1986-96), the tendency for women's self-employment to increase seems clear in a number of countries. In Australia and in the United States in particular, however, the progress in female self-employment has been significant and steady over both of the last two decades. But self-employment data is only one aspect of the creation and management of women-owned businesses.

In the United States (...) in the last few years the number of firms created and managed by women has grown twice as fast as those set up and managed by men.

With regard to these businesses in the United States, for instance, where research is most developed, the number of firms created and managed by women has grown twice as fast as those set up and managed by men in the last ten years. Approximately eight million businesses are owned and managed by women. According to the National Foundation for Women Business Owners (NFWBO), firms headed by a woman provide one out of every four private sector jobs.

Women-owned enterprises in the United States, 1987-96

	1987	1992	1996	Variation 1996/1987	Average annual change
				Percentages	
Number of enterprises (millions)	4.47	6.41	7.95	+ 77.6	+ 6.6
Employment (millions)	6.55	13.21	18.54	+ 182.7	+ 12.2
Turnover (billions)	681	1 574	2 288	+ 235.8	+ 14.5

Source: NFWBO (1997).

There are many indications of the growing importance of women-owned SMEs in OECD and non-OECD countries

Despite the scarcity of internationally comparable data on women-owned SMEs, a significant number of recent observations either demonstrate the substantial importance of women-owned SMEs at a particular moment in time, or show an increase in the number of women entrepreneurs. For instance, **Shigeko Mitsuhasi [Japan, Access to Markets]** indicated that the percentage of women entrepreneurs in Japan increased from 2.4 per cent in 1980 to 5.2 per cent in 1995 (see table below).

Women as a percentage of entrepreneurs in Japan, 1980-95

	1980	1982	1984	1986	1988	1990	1992	1994	1995
Number of women	11 612	17 553	23 021	28 007	34 836	41 171	46 958	50 712	51 911
%	2.4	3.2	3.7	4.0	4.4	4.7	4.9	5.2	5.2

Source: Nikkei Research Institute of Industry and Markets.[1]

In France, a 1990 study by the *Crédit d'Équipement des Petites et Moyennes Entreprises* (CEPME) showed that in enterprises employing between three and 500 employees (of which there are about 550 000 in France), women were running 19 per cent of firms that were less than ten years old, as opposed to 15 per cent of businesses created more than ten years earlier **[Henry Savajol, France, Financing]**. **Lorraine Martin [Australia, Business Environment and Technology]** pointed out that, in 1990, women's small businesses were estimated to be responsible for approximately 10 to 15 per cent of Australia's national output, and provided about half of the country's private sector employment. **Maria Teresa d'Avila [Portugal, Financing]** indicated that according to the Portuguese Association of Women Entrepreneurs (AMEP), there are approximately 380 000 women entrepreneurs in Portugal. She also stressed that a cross-national study carried out in the United States reviewing some 30 industrialised and developing countries, showed the growth of women entrepreneurs to be fastest in Portugal (153 per cent) between 1991 and 1995.

In non-OECD countries, growth in women-owned businesses is also apparent. It would appear that the growing legitimacy of women's participation at high levels in family-owned firms in Asia and Latin America is contributing to this trend. **Rajni Aggarwal [India, First Plenary Session]** mentioned that the 1991 census in India showed women entrepreneurs to represent 10 per cent of all companies. She also emphasized that by the year 2000, this figure could rise to 20 per cent.

According to the United Nations, in the Latin American universe of micro/small businesses, the female segment may compose nearly half of these firms, when both the formal and informal sectors are included **[Sonia de Avelar, Brazil, Dialogue among OECD Countries, Transition and Developing Economies: Future Challenges for Policy Makers** (hereinafter referred to as **Dialogue...)]**.

Women entrepreneurs clearly make an important contribution in other developing countries. **Leyla Khaïat [Tunisia, Management]** stressed that in Tunisia, for example, women are playing an increasing role in the country's development. These trends are inadequately documented, however. Women are known to be intensively involved in agricultural production; they are estimated to be responsible for 80 per cent of food production in Africa, 60 per cent in Asia and 40 per cent in Latin America. They are also responsible for commercialising food – providing them with excellent knowledge of local markets and trends.

Women in the transition economies are increasingly turning to entrepreneurship.

Statistical data broken down by gender are scarce in the transition economies, but evidence there suggests that women, who are particularly affected by the privatisation of state-owned industries and the structural unemployment resulting from the transition to market economies, are increasingly turning to entrepreneurship, albeit from very low levels. In Romania, for example, women create large numbers of SMEs, either alone or in association with others **[Raluca Bejan, Romania, Dialogue...]**.

What is unique about female entrepreneurship?

"What is important about women's entrepreneurship? Why is women's entrepreneurship different? How do the differences matter? Why are the differences important and what needs to change? World Bank and UN research, as well as empirical studies across countries, make it fairly clear that occupational segregation exists in almost every OECD country and in many of the less developed countries. Women do not have equal opportunities in terms of work experience. They come to entrepreneurship with different sets of work experience, and therefore create different kinds of ventures. They do not have the same work experience in the same sectors as their male counterparts, which makes it harder in some ways for them to create businesses in sectors where they have no experience. Social expectations regarding women's role and family responsibilities usually mean that women assume a greater responsibility for household, childcare and dependent care which can be a burden for women trying to manage and balance these responsibilities while trying to grow their businesses. Organised social life: in some countries women are less welcome in social networks (for instance going out after work for cocktails) and are left out of some of those loops, meaning they do not have access to as much information. So social structures and the way that women socialise influence the human and social capital endowments with which they start their businesses." **[Candida Brush, United States, Management]**

Recent improvements notwithstanding, women business owners receive insufficient attention from national and local authorities, educational institutions and the world of business and finance who do not yet take their role in the economy seriously. Despite the scarcity of systematic data, certain general observations can be made regarding some of the specificities of women-owned businesses and the obstacles they face:

The great majority of businesses run by women are in the service sectors.

◊ In all OECD countries, the great majority of businesses run by women are in the service sectors. A poll of 17 000 women in several European countries (with the exception of Spain and Portugal) showed that almost 5 million women work independently. Almost 46 per cent of these women are in retail, 12 per cent in the beauty and skin-care business, 10 per cent in professional services (doctors, lawyers, etc.), 9 per cent in crafts, and only 1 per cent in the manufacturing sector.

Distribution of women-owned firms in the United States is very much skewed to the small end.

◊ In the United States about 75 per cent of women-owned firms have no employees, a figure similar to that for all US businesses; fewer than 1 per cent of women-owned firms have revenues of US$1 million or more, while about 2 per cent of all US firms have revenues of more than US$1 million. Firm distribution is very much skewed to the small end; women have slightly smaller businesses and are slightly younger; in firms with employees, about half have fewer than ten members of staff **[Candida Brush, United States, Management]**.

◊ Evidence suggests that there is a close relationship between obstacles and type of business – women in enterprises engaged in activities considered to be extensions of the domestic arena face fewer problems **[Gina Zabludovsky, Mexico, Management]**.

◊ In some countries, female businesses seem to have a higher rate of failure than male businesses, which may have to do chiefly with the type of business, although it may also reflect the obstacles faced by women. In other countries, the reverse is true, a trend which has been linked to women entrepreneurs' more risk-averse attitudes. Research conducted in the United States (NFWBO) and Canada, found that three out of four female-owned companies stay in business longer than three years, compared to two out of three male-owned companies.

Access to capital is a primary barrier encountered by women entrepreneurs..

◊ Access to capital is a primary barrier encountered by women entrepreneurs. This theme was addressed in one of the conference workshops (see below).

◊ The obstacles to visibility and recognition that women face are probably exacerbated in developing countries where women are not easily accepted as business owners. These obstacles affect women's access to capital and make it more difficult for women entrepreneurs to acquire physical assets for a business, rent facilities, or even hire employees.

Recommendations:

The OECD conference proclaimed clearly and repeatedly that policy makers face a real challenge in finding appropriate ways to encourage the collection and standardization of data at national and international levels, and to broaden our understanding of women's entrepreneurship in order to increase the overall effectiveness of SME policies.

What data and information should be collected nationally and internationally?

Basic information from around the world is required, and further study needs to be made even in countries, like the United States, where the basic information is known.

The general view is that basic information from around the world is needed. Even in the United States, where the basic information is known, "Further study needs to be made," as **Julie Weeks [United States, Business Environment and Technology]** pointed out. In all countries, data is needed to answer questions such as:

◊ How many women-owned businesses are created, and in which industries? It is important to highlight "non-traditional" industries for women, such as construction and manufacturing *versus* the service sector.

◊ Which businesses are growing, and at what rates?

◊ How can women business owners be described and classified?

◊ How do start-ups compare to firms that are over ten years old?

◊ What happens as women-owned businesses grow and become more established?

◊ How do different management styles between men and women affect business growth and profitability?

Data collection should be modified to permit statistical breakdowns by gender for SMEs involved in international trade.

Virginia Littlejohn, David Bowie (United States) and Angeline Low (Australia) [Globalisation] suggested that existing data collection should be modified to permit statistical breakdowns by gender for SMEs involved in international trade. Workshop 5 on Financing Issues also recommended that data related to the financing of SMEs be collected on a gender-specific basis. The importance of evaluation of programmes in order to gain a better understanding of successful practices was also reiterated.

How and by whom?

Governments have a role to play in improving data collection and information on women-owned businesses and in deepening our understanding of these businesses.

Participants emphasized the role of governments in improving data collection and information on women-owned businesses and in deepening the knowledge of these businesses. However, the need for a joint effort between governments, associations, private corporations and research organisations to foster this type of work was also repeatedly stressed. The creation of an international network of governments, corporations and women-owned businesses world-wide was suggested (Workshop 2 on Globalisation and International Trade).

Technology can facilitate the task of collecting and compiling data.

The use of technology is a means of facilitating these tasks. Existing data or means of compiling data, (*e.g.* census data and NFWBO research) should be used, and existing research better shared.

A role for the OECD.

The OECD should play a leadership role in anchoring and facilitating the collection and sharing of women-related SME data on a world-wide basis.

ANNEX 1

Share of women employers and persons working on own account in selected OECD countries: 1976, 1986, 1996

(excluding the farm sector and family workers)

Country	1976	1986	1996	Average annual change 1976-86	1986-96
Australia	26.0	30.9	32.8	0.49	0.19
Belgium[1]	28.3	28.4	28.7	0.01	0.03
Canada	n.a.	n.a.	41.6	n.a.	n.a.
Czech Republic	n.a.	n.a.	27.4	n.a.	n.a.
Finland[2]	34.8	36.1	30.5	0.14	-0.56
Germany[3]	34.3	26.6	27.9	-0.77	0.13
Greece[4]	18.2	16.4	19.7	-0.20	0.37
Hungary	n.a.	n.a.	30.7	n.a.	n.a.
Italy	27.4	22.4	23.7	-0.50	0.13
Japan	33.0	35.9	32.2	0.29	-0.37
Korea	n.a.	n.a.	30.7	n.a.	n.a.
Mexico[5]	n.a.	n.a.	36.1	n.a.	n.a.
Norway	23.9	23.4	30.3	-0.05	0.69
Poland	n.a.	n.a.	30.7	n.a.	n.a.
Spain	26.2	24.8	27.7	-0.14	0.29
Sweden	22.6	31.2	25.9	0.86	-0.53
Turkey	n.a.	n.a.	5.3	n.a.	n.a.
United Kingdom[6]	21.0	25.7	30.6	0.47	0.61
United States[7]	26.9	33.3	39.2	0.64	0.59

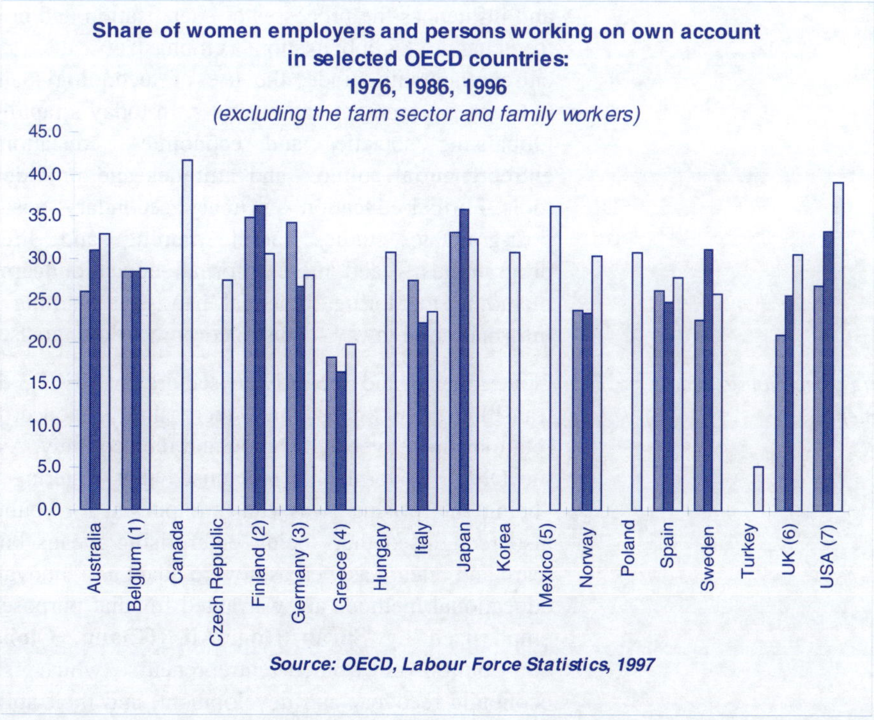

1. Last value corresponds to 1992. **2.** First value corresponds to 1977. **3.** From 1991, data refer to the unified Germany. Prior to 1991 data refer to Western Germany. **4.** First value corresponds to 1977 and last value to 1995. **5.** Last value corresponds to 1994. **6.** Due to changes in census data from 1990 onwards are not directly comparable with those of previous years. **7.** Value corresponds to 1995.

Source: OECD (1997*a*), *Labour Force Statistics.*

B. PROMOTING ENTREPRENEURIAL EDUCATION AND TRAINING

Key Issues

Why should there be a special focus on increasing the entrepreneurial know-how and capabilities of young girls and women? Do some groups of boys and men also benefit? Does this benefit society and the economy as a whole? Who has key responsibilities in fostering an entrepreneurial transfer of knowledge? How can education prepare and encourage male and female students to become entrepreneurs? When should basic business literacy be acquired and by whom? What can learning institutions do to adopt methods at all levels which focus on personal development for all? These are some of the questions and issues debated during the conference sessions.

"Stereotypes must change, and an entrepreneurial spirit must be fostered in all schools so that we are less passive, less paternalistic, and present better role models. We must adopt non-traditional, inexpensive, widely available training methods, and deliver them to a large segment of the population, especially to women in rural and depressed areas to help them find business opportunities, work at home, or study. Manuals, literature and motivational programmes should be used to promote entrepreneurial attitudes generally – and not only in education. Innovative methods (easy-to-read magazines, comic books, videos, cassettes, books, multimedia, and Internet) could help to encourage women to think about starting a business. Advertising campaigns could help change the traditional perception of women as passive and non-competitive. Modern, culturally and socially active women in professional and decision-making roles can help spark an entrepreneurial attitude in the public, particularly among the younger generation." **[Salo Grabinsky, Mexico, Management]**

Education plays a key role in providing skills and know-how ...

Education plays a key role in sustainable economic development. It provides the skills, knowledge and know-how needed to stimulate the innovative potential of active, contributing economic agents, and influences the processes of socialisation and how people adapt to change. Globalisation emphasizes the private sector, entrepreneurship, trade, the use of technology, and partnerships between government and business. In today's rapidly evolving and globalising societies and economies, education must foster entrepreneurial cultures and attitudes and provide the necessary tools. Formal education – primary, secondary, post-secondary and post-graduate studies, adult training and lifelong learning programmes – and more informal education/learning systems – including mentoring, practical hands-on learning and using the information highway – must incorporate these goals.

... contributing to economic recovery and development ...

Governments and the private sectors in OECD, developing and transition economies are increasingly concerned with the relationship between education and the economy. As early as 1988, the OECD governments recognised that fostering competence in "being enterprising" was vitally important for young people. They also recognised that "being enterprising means knowing how to learn and relearn as well as how to adapt and innovate". Changes in educational methods are warranted for that purpose.[2] In Africa, as emphasized by **Sarah Hage-Ali [Ghana, Globalisation]**, the contribution of women entrepreneurs, which is important to economic recovery and development, and their ability to compete in the global economy are linked to the enhancement of education for women.

... and fostering an entrepreneurial

Appropriate education and adult training are key to fostering widespread entrepreneurial spirit. Education can create new *culture.*

entrepreneurs who can innovatively use and acquire knowledge, encourage self-employment, new businesses, and job creation. Understanding how to learn from the multitude of information sources available is also essential to ensuring the competitiveness of firms.

Technology is an important driver of economic growth: women and girls should be encouraged to acquire skills in the "hard sciences".

The creators of technological innovations which transform businesses and keep them competitive are key to the flexibility needed in today's economies. In many OECD countries, however, "male students constitute the majority in technology, mathematics, engineering, and computer science programmes (which) do not seem to be conducive to female students; discrimination against those women who do insert themselves in these areas is quite widespread. Technology is an important driver of economic growth: women and girls should be encouraged to acquire skills in the 'hard' sciences" **[Amy Millman, United States, Business Environment and Technology]**.

Policy Implications and Good Practices

Government-business partnerships

Education systems need to be adapted to current needs, providing effective programmes for men and women alike.

Governments tend to play a key role in shaping formal education systems. Business plays an increasingly important role in shaping adult-training and more informal education systems. Governments, learning establishments, and other intermediary structures should work in partnership with micro, small and medium-sized business owners, both men and women, to ensure that formal and informal education systems are usefully adapted to current needs, and designed to address these different stages of growth. They should also ensure that programmes reach men and women, and are effective for both. In Australia, the State of Queensland has a programme, within the curricula of enterprise training and government-provided assistance and funding, for real businesses to come together with high school students and create real enterprises **[Judith Maestracci, Australia, Access to Markets]**.

Adapting education systems to the client

Education needs to be flexible enought to adapt to shifting demand ...

Education systems must be better adapted to specific circumstances. They should use a mix of classical and non-traditional teaching methods and approaches, and distinguish between the needs and competence of sub-groups, including those doing business in urban or rural areas, those providing goods or those providing services, firms run by men or those run by women, start-ups or expanding firms. Education and training systems and technology must be flexible enough to adapt to shifting demand. They must include a comprehensive approach to lifelong learning.[3]

... and ensure that men and women achieve comparable results.

Opinion divides on whether education and training programmes related to building a capacity for entrepreneurship should be gender-specific and designed specifically for boys/girls, and men/women. Socio-political, financial and cultural circumstances tend to determine which approach is more effective. In all cases, formal and informal education systems should ensure that women and men achieve comparable results and that all students participate and benefit equally.

Men and women learn differently for many reasons, both cultural and historical. Girls and women seem to learn differently in single-sex and mixed-sex classes, and seem to respond positively to "tailored" methods. Empirical evidence seems to suggest that teaching methods used to ensure female participation in a co-educational class can also benefit boys.

Women should be promoted as role models for aspiring entrepreneurs.

Methods for teaching adults about entrepreneurial skills, however, tend to disregard gender differences and provide male-oriented case studies and roles. Wherever entrepreneurial competence learning is provided, the materials should feature women as decision makers and as models for aspiring entrepreneurs. Women often have different experience, needs and preferences, have made different choices and pursued different tracks, have left their jobs and spent several years caring for children and husbands before reintegrating the formal economy by starting a business. They are often more interested than men in management skills and issues, and less interested in finance. Experience in Australia, Canada and the United States suggests that women tend to prefer less formal, experienced-based training; to learn from women; to be helped/mentored by other women; and to receive follow-up training. Women especially need to learn about business finance and technology, as well as business relationships and successful strategic alliance building.

Evidence suggests that women learn better when they learn from women...

As **Candida Brush [United States, Management]** emphasized, women business owners manage businesses differently, and their management techniques should be used to teach different ways of looking at effective management practices. In the United States, the National Association of Women Business Owners (NAWBO) was formed in 1975 and represents nearly 8 million women business owners. It helped lobby to create the Office of Women's Business Ownerships, known as OWBO, which is housed at the US Small Business Administration [**Suzanne Taylor, United States, Business Environment and Technology**]. "OWBO has 65 women's demonstration sites which provide training and technical assistance to women entrepreneurs and will soon launch on the Internet the first ever interactive business skills training site for American women entrepreneurs. Technology is used as a tool to train women, especially in areas that are difficult to access or to grasp." **[Sherrye Henry, United States, Business Environment and Technology]**.

... and that mentoring is an effective teaching method.

Mentoring is an effective teaching method, and women business owners effectively mentor their peers. Australia introduced the Step-Ahead Business Mentoring Programme in 1995 and has established a pool of women and men to act as role models for other business women. This has enhanced the growth and visibility of women-owned businesses; almost all of the 2 000 women participants increased their personal and business skills and confidence **[Angeline Low, Australia, Globalisation]**. The importance of role models was also stressed by **Zenaida Gordon [Philippines, Dialogue...]**: "Elena Lim is our leading woman entrepreneur in the Philippines. She is a symbol, a role model for all of us, because she has made it happen; she has entered into joint ventures with AIWA, with Sony, with Samsung, with Sanyo".

Education and training systems should encourage women to use information techniques and information technology more systematically.

Informal settings also serve to transmit useful information for business. Networks, associations and relationship building are key to keeping abreast of techniques, business information and opportunities. Women use networks and associations less than men do because they have relatively less time to devote to networking, and have been discouraged by their inability to break into "boys' clubs". Where useful, business information deliveries should distinguish between women- and men-owned businesses. Education and training systems should encourage women to use information techniques and information technology more systematically.

> "It is a priority to promote successful role models and to make those examples of success more visible to young women and to the public at large, so that there is more openness to the possibilities that women not only *can* do these things, but *are* doing these things, and are doing these things very successfully." **[Amy Millman, United States, Business Environment and Technology]**

Teaching entrepreneurial skills to children

Entrepreneurial skills more effective the earlier they are passed on.

Children whose parents run their own business, or are "entrepreneurial", are most likely to develop entrepreneurial attitudes. Practice has shown that the earlier basic skills are passed on, the more effective they are. The President of the US Export Council has developed a multimedia programme called "Virtual Trade Mission" which teaches secondary or high school students and undergraduates about the importance of exporting. An "Income of Her Own" teaches young women about entrepreneurship using real case studies. The United States has summer camps where young girls learn to operate a business **[David Bowie, United States, Globalisation]**. In some schools eight- and nine-year olds learn to handle "virtual" stock portfolios and simple businesses.

Children benefit from visits to their mothers' places of work (...) reinforcing the positive role model of the working mother.

Children better understand businesses and how they are run after visits to manufacturing or technology firms **[Dina Lavoie, Globalisation]**. The United States has launched a "take your daughter to work day" **[Cherie Piebes, Business Environment and Technology]** which reinforces the role of the working mother as a positive role model to emulate and to instil qualities of activity, independence and enterpreneurship. Women need to learn that other women have families, marriages and run businesses at the same time.

Encouraging entrepreneurial competence in children and adults, and in formal and informal educational contexts, requires ensuring familiarity with basic economic, business and financial concepts, management skills, and information technologies.

Affordable and accessible learning

Easily accessible education, using computer technologies and hands-on training, works best.

Knowledge and competence must be accessible, inexpensive and not overly time consuming. Women tend to own small businesses with slim profit margins, and to have less time than men to run them. They are more likely to use inexpensive and time-efficient access to knowledge and competence building. Moreover, easily accessible education, available at a range of sites, will be most effective: educational centres and institutions; at home through Internet and computer technology; within medium-sized businesses, especially via hands-on training; corporate mentoring programmes. The US government has created 53 centres where women can acquire business skills training **[Sherrye Henry, United States, Business Environment and Technology]**.

Technology as a key learning tool for firms

New technologies can ensure that teaching materials are as user- and learner-friendly as possible.

Education teaches how to obtain pertinent and useful information and knowledge. Global information networks which provide electronic training programmes, share information, best practices, case studies, and market and business information are good examples and are part of the ongoing learning that entrepreneurs require in order to remain competitive.

The technological boom, the rapid expansion of the Internet and other innovative technological, audio-visual and computer tools should be used to ensure that teaching materials are as user- and learner-friendly as possible.

Recommendations:

A number of proposals and recommendations relevant to education and training emerged from the various workshops and plenary sessions of the conference:

- ◊ *Revisit teacher-education curricula to include basic entrepreneurial skills. Revisit teaching curricula for children and adults to ensure that basic business, management, finance and technology skills are addressed. Establish policy to ensure that teaching materials are gender balanced.*
- ◊ *Governments can help establish learning partnerships that promote "experience-exchange workshops", training programmes (especially in marketing, finance and technology), gender exposure visits and eye-opener activities. These partnerships would be the key to the success of any national and international partnerships between education institutions and the business community, in OECD Member and non-member countries alike.*
- ◊ *Encourage the establishment of centres with "volunteer" businesses to provide education and training facilities, material, programmes, information. Ensure their accessibility and affordability to women. Encourage business to recycle "old" equipment, especially computers, to advisory and learning centres.*
- ◊ *Evaluate training programmes for cost-benefit analysis data and see how effectively enterprise education is working for young women and girls in schools.*
- ◊ *Initiate campaigns to promote female entrepreneurship.*
- ◊ *Collect statistics on the use and effectiveness of "women-only" education and training programmes for entrepreneurial skills.*
- ◊ *Encourage the creation of a global network for providing and sharing information, best practices, case studies and other educational and training material. Ensure the network's accessibility to a maximum number of users.*
- ◊ *The OECD should disseminate good practices related to training programmes and entrepreneurial education for girls and women. The OECD should also disseminate good practices based on successful learning partnerships between education institutions and the business communities in OECD, transition and developing economies.*

C. Removing Financial Obstacles and Promoting Innovative Tools to Foster the Creation and Growth of Women-owned SMEs

Key Issues

SME financing

The conference sought to highlight issues related to financing women-owned businesses and how policy and market-led developments can improve their access to financing.

> "We need a strategy that is multi-faceted: a gender-specific set of strategies and a mainstream strategy. Government must first exert leadership, be a catalyst, a problem solver and a risk sharer. Next, we must identify potential partners in collaborative investment opportunities. These can include business mentors, mainstream investors, and financial intermediary organisations. Third, we must create and implement new interventions and financial products to elevate women's business investments into a comprehensive financial system. Last, we must employ the new telecommunication technology as a channel for matching investors and women business owners. Women leaders must engage in shaping public policy for the full spectrum and full continuum of financing, from the smallest to the largest business." **[Linda Tarr-Whelan, Financing]**

Are the difficulties that women encounter specific to their status as women, or are they based on the types of SMEs that they set up and the sectors of activity in which they choose to do business?

SME financing is much discussed by business managers, bankers, public authorities, and domestic and international economic organisations. Women business owners share basic characteristics with their male counterparts and must contend with an increasingly demanding financial environment. Access to capital, and particularly to start-up capital for women, is made very difficult by the hurdles posed by guarantees, the cost of money and loan ceilings. Are the difficulties that women encounter specific to their status as women, or are they based on the types of SMEs that they set up and the sectors of activity in which they choose to do business?

It would appear that the perceived disparity and lack of credibility cited by many women entrepreneurs stem primarily from banks' wariness of certain lines of business, a certain size of enterprise, and a lack of experience in assessing initial needs and financial planning.

Patricia Mahot [France, Financing] emphasized that few entrepreneurs obtain bank loans at the outset, and those who do include as many women as men. The *Observatoire des Entreprises*'s 1995 panel study of 30 000 French entrepreneurs showed that 23 per cent of female entrepreneurs had received bank loans, compared with 22 per cent of male entrepreneurs. Upon analysis, it would appear that the perceived disparity and lack of credibility cited by many women entrepreneurs stem primarily from banks' wariness of certain lines of business, a certain size of enterprise, and a lack of experience in assessing initial needs and financial planning. In any event, these figures show that 75 per cent of all entrepreneurs had no bank loans for their start-up phase. These data require further refinement, however, because they can mean bank refusals, or underestimated or non-existent needs. The Austrian Institute for Small Business Research published a study in 1996 that concluded, similarly, that the greatest problem for 60 per cent of men and women entrepreneurs alike was access to financing.

A study in Sweden and another in Switzerland show that men and women in the same line of business have more in common than do two women working in different sectors. All entrepreneurs face some fundamental financial problems, but these can be more difficult for women SME owners to resolve.

Because most SME owners, or future owners, have limited personal funds, they must often borrow both to set up and to develop their companies. Most frequently, creditors have imperfect information about the quality or intentions of potential borrowers, which creates contract problems. The creditor must gather information about the debtor and establish a sense of trust. However, this requires a sunk, fixed cost that grows in inverse proportion to the information that the creditor has about the debtor and the consonance of his/her experience or approach to management. Transaction costs are also disproportionate to the small loans requested. Small entrepreneurs are perceived as lacking visibility, having few prospects for growth and being in sectors where the entrepreneur is the company's main asset. As a result, risk evaluation is often seen as a gamble for the bank.

Funding for women-owned enterprises

However, women do encounter particular difficulties in accumulating start-up capital and access to financing.

Notwithstanding the problems common to both sexes, women encounter particular difficulties in accumulating start-up capital and accessing financing:

◊ Most women establish very small enterprises in the service and retail sectors, the less expanded sectors, and generally operate in local markets. In many countries, enterprises operating in the retail trade sector tend to have a much lower survival rate.

◊ Women tend to gather less start-up capital than men as a result of difficulties in saving, linked to lower personal assets or former incomes as employees.

◊ Women typically apply for smaller loans. Bankers perceive these as akin to personal loans. Underestimation of their enterprise's financial requirements or a reluctance to borrow larger sums can adversely affect the perception of banks and organisations of firm viability. **Linda Tarr-Whelan [United States, Financing]** mentioned a recent US Federal Reserve study of small-business financing that indicates that 42 per cent of all women-owned firms used their personal credit cards for short-term financing, and 30 per cent financed their business from personal savings; 42 per cent of women-owned businesses have assets of less than US$25 000; and more than 20 per cent of women-owned businesses have annual sales of less than US$25 000.

◊ Regarding guarantees, self-employed women, or women business owners operating under specific juridical statutes, are often reluctant to use family assets as collateral. In some cases they must procure their husband's or family's agreement to obtain loans. Notwithstanding discriminatory bank regulations in certain countries (unequal inheritance laws, treatment of women with regard to a couple's common assets in the event of divorce, etc.), the status of women co-entrepreneurs who are part of a family business does not always reflect their entrepreneurial contribution (and can result in lost retirement or social security benefits).

◊ Women often have more acute information and public relations problems. **Fotini Legaki [Greece, Financing]** pointed out that it is difficult for a woman to forge social ties with bankers and to establish a strong personal relationship, that SME owners are often ill-informed about new financing tools such as leasing, factoring, venture capital, etc. **Dina Lavoie [Canada, Financing]** stressed that few women benefit from institutional grant programmes, often because they receive the information too late – "they are not on the mailing list". Women are seldom asked to be on the boards of development banks that make business loans. The bureaucratic process involved with each bank transaction is a significant time constraint for a small firm. **Naoko Banno [Japan, Management]** reported that her company did not receive a grant to develop a software programme because of the difficulties raised by the negotiation and bureaucratic process.

◊ Because women have entered the labour market later than men in some professions, they are often less experienced professionally, particularly in management, supervision and access to finance.

"We should first do all what can be done to raise women's income. Access to bank finance is important but equally important is the ability to accumulate capital." **[Leila Webster, The World Bank]**

Policy Implications and Good Practices

Schemes providing financing for micro enterprises and very small firms can be expected to facilitate the task of a large number of women entrepreneurs.

It is important to examine the existence and relevance of market responses against this background and, in the case of market failures, of government programmes aimed at facilitating the access of women entrepreneurs to financing. Evidence suggests that some financial organisations have developed lines of credit for women-owned SMEs and that there are some success stories. Indeed, **François Beaudoin [Canada, Financing]** reported that "50 per cent of new businesses, and often micro businesses, are actually started by women. The Business Development Bank of Canada (BDC) forecast that within the next ten years, the proportion of businesses owned by women and men will be about equal in Canada, given the rate at which women entrepreneurs are starting new businesses. This is an important market for any financial institution."

The various schemes that can provide satisfactory responses for obtaining financing for micro enterprises and very small firms can therefore be expected to facilitate the task of a large number of women entrepreneurs.

Other schemes seek to bolster equity by bridging the gap between financing by friends and relatives and investment by venture capitalists.

Henry Savajol [France, Financing] reported on French schemes seeking to bolster equity by bridging the gap between financing by friends and relatives and investment by venture capitalists. *La Banque du Développement des PME* (Development Bank for SMEs) has recently proposed measures to encourage direct individual investment in unlisted SMEs through specialised investment funds, eligible for partial state guarantees, through one of its specialised institutions, SOFARIS (*Société française de garantie de financement des PME*). Similar possibilities already exist for venture capital firms, and to develop procedures for quasi-equity investments related to small businesses in the development stage. This technique is inspired by the equity loans developed in France in the 1980s and by the "venture loans" of the Business Development Bank of Canada. These are unsecured six-year loans with a low base rate of interest and with royalties based on the borrower's turnover.

Guarantee schemes provide security to facilitate bank lending ...

Henry Savajol also emphasized that to facilitate access to bank financing for micro enterprises with few prospects for rapid growth, banks must have the security that entrepreneurs themselves cannot provide. Guarantee schemes are perfectly suited to such situations. In France, there are a number of different types of guarantee schemes: *e.g.* private funds (mutual guarantee organisations); public funds administered by SOFARIS; and a special guarantee fund for women (FGIF, created in 1989) who are involved in creating, acquiring or developing a business (loans of under FF 150 000 are eligible; and guarantees cover 60 per cent of the loan).

Nancy Barry [United States, Dialogue...] stated that Women's World Banking has been successful in developing the concept of capitalisation as grant equity for micro-financing institutions in the developing countries. This means being able to build their own portfolios, and to lend and then to leverage that money for commercial funds in the local banking system. A similar concept can be developed for business development service organisations to improve access to financing for women, with a view to providing solutions to broad-based economic development.

... while other programmes allow women to obtain a letter of pre-qualification before presenting their loan proposal to the bank...

Sherrye Henry [United States, Business Environment and Technology] explained that the Small Business Administration (SBA) has started a micro-loan programme which offers women smaller loans (US$500, 5 000 and 25 000). The loan documentation programme also offers women the possibility of the SBA guaranteeing their bank loans up to US$100 000. The SBA also provides a women's qualification loan programme which allows women, particularly in service businesses, to obtain a letter of pre-qualification before presenting their loan proposal to the bank.

... and yet others combine loans, training and counselling.

Other programmes, such as those of the BDC, combine loans, training and counselling. **François Beaudoin [Canada, Financing]** explained that three years ago, the bank launched the Step-up Programme, a mentoring programme whereby women entrepreneurs starting a business meet more experienced entrepreneurs. Since its inception, there have been 592 successful mentoring assignments; there is a strong link between improvement of management skills and financial performance. Mr. Beaudoin also stressed that 30 per cent of the BDC loan portfolio is given to women entrepreneurs and to joint ownership where at least one of the majority shareholders is a woman. Bank experience proves that loan losses are lower with women entrepreneurs. The BDC has also developed a micro-business programme offering training, counselling and financing. The bank considers that the initial phase of this programme has been successful: during a 12-month period (1996-97), 384 loans were approved for a total of C$ 12 million. At this point, only 30 per cent of the participants are women entrepreneurs, but this percentage is expected to increase substantially. Of the women entrepreneurs who followed the micro-business programme, 32 per cent are involved in knowledge-based and export businesses, 22 per cent in trade and 18 per cent in manufacturing.

Some banking institutions lend to new businesses that do not qualify for conventional loans ...

Four years ago, a leading Dutch bank, the ABN AMRO Bank, set up a department for business founders of which 20-25 per cent are women. According to the bank, women receive loans that are almost two-thirds smaller than those provided to men because women-owned start-ups tend to be part-time businesses that necessarily have lower turnover and funding requirements. **Robert Hughes [Ireland, Financing]** gave the example of the Bank of Ireland which has a fund of some 60 million ECU for start-up loans to new businesses that fail to qualify for conventional bank lending. The bank also plays an active role with the First Step programme, which provides financing and counselling for small businesses in the start-up phase.

... whereas the Loan Support Scheme for Women in Sweden is a new concept tailored by the state to support women entrepreneurs, both in new businesses and ongoing firms ...

Other public support programmes also exist at national and international levels. **Anne Löfquist [Sweden, Financing]** reported that Sweden's Loan Support Scheme for Women is a new concept tailored by the state to support women entrepreneurs, both in new businesses and ongoing firms. In addition to a loan which can amount to 15 per cent of the capital needed (and can range between 3 000 and 17 000 ECU), the programme includes mentoring projects, training, seminars, training for board members, etc. Since its beginnings in 1994, 5 600 new jobs have been created and over 3 000 firms have borrowed more than 26 million ECU. Forty per cent of the loans are in the service sector, 50 per cent in trade (retailers), and only 8 per cent are in manufacturing. The duration of the loan is six or nine years; the first two or three years are free of interest and amortisation. Ms. Löfquist stressed that the support is considered to have been an important incentive for women to start or expand their own business.

... and the European Commission has been dealing with business creation by women, including the financial aspects, for over ten years.

Hella Gerth [European Commission, Financing] explained that the European Commission has been dealing with business creation by women since 1986 in its programmes related to employment creation. The Local Employment Initiative and later the New Opportunities for Women Programme have had good results and have created positive examples of women's entrepreneurship. These examples have been evaluated and the New Opportunities for Women Programme will continue until 1999. The gender perspective has been integrated into the main support programmes of the European Community. The European Commission has a small budget for pilot actions, seminars, and conferences for all small/micro entrepreneurs, men and women alike.

The Grameen Bank was established in Bangladesh by Dr. Yunus in 1983. Dr. Yunus revolutionised lending to the very poor by developing a "technology" based on powerful repayment incentives (group loans with mutual repayment obligations) and prompt, attentive follow-up with local supervision and repeat loans. As a result of this pioneering work, more than 20 million people, almost exclusively women (94 per cent), have been able to break out of extreme poverty through micro-entrepreneurship. In Bangladesh 10 per cent of the population (10 million people) benefit from the Grameen Bank. The Grameen Bank model, with its 90 per cent repayment rate, has been successfully repeated and integrated into the private and public lending schemes of more than 56 OECD and developing countries, including Canada, China, Norway, South Africa and the United States (Arkansas and Chicago).

Recommendations:

It is important to deepen our understanding of women-owned businesses, particularly very small enterprises (financing arrangements for start-ups, personal savings, loans from relatives, bank loans, average loan size, percentage of successful loan applications, percentage of failed loans, etc.). It is also important to assess the relevance and effectiveness of "good practices". Are they "best practices"? Should they be disseminated so as to promote proven financial products and tools?

The debate around these issues was controversial, in particular whether to promote specific actions, schemes or support measures. "Does specific programming for women represent discrimination, or is it positive action? We need positive action to redress imbalances, but every country now has an equality law, an anti-discriminatory law. The Loan Support Scheme for Women initiated in Sweden would not be accepted in many countries" ***[Robert Hughes, Ireland, Financing]****. "My plea would be, no sex discrimination in these recommendations, because these recommendations are just as important to male entrepreneurs as they are to female entrepreneurs"* ***[Ian Hamilton Fazey, United Kingdom, Financing]****. However, it was generally agreed that finding solutions for women-owned businesses would benefit small firms in general.*

Conference participants also recognised the importance of trying to match the objectives of providers of capital, in terms of investment criteria and capital needs, with the needs of small businesses to make their businesses thrive at different stages in the life-cycle. This could result in identifying currently missing partners able to fill some gaps.

Participants agreed that in order to facilitate access to, and a sound continuum of, credit to micro enterprises in general, and women-owned businesses in particular, the perceived "risk" of lenders needed to decrease. The following approaches might be considered by policy and decision makers and adapted at national and local levels:

- ◊ *Promote greater women business owners' equity by increasing women's assets through savings programmes, home ownership, etc.;*
- ◊ *Promote appropriate loan guarantee schemes;*
- ◊ *Make a credit pool available to women from the smallest to the largest business opportunity in a "secondary market" for small business loans; develop micro-business financing programmes;*
- ◊ *Increase access to credit by promoting the development of intermediary organisations, intermediary structures between large pools of capital and very small borrowers (Grameen Bank) which can deliver small amounts of money to entrepreneurs at reasonable cost;*
- ◊ *Increase women's business success by promoting counselling services to women business owners from start-up to development, and access to support networks for training, information, market development, as well as social protection (health insurance) and child care;*
- ◊ *Promote possible partnerships between government and NGOs, and private sector lending in order to make this market attractive;*
- ◊ *Encourage international organisations, such as the OECD, to further the knowledge about women-owned business financing and disseminate best practices in this field in OECD, transition and developing countries.*

D. Promoting the Competitiveness of Women-owned SMEs

Key Issues

A world dominated by knowledge, networks and interactive communication, new technologies, innovation and social concerns (...) and a growing number of women entrepreneurs.

The world is changing and the business environment is evolving rapidly. Futurists predict that the 21st century will be dominated by knowledge, networks and interactive communication, new technologies, innovation and social concerns (such as the continuing need to maintain living standards and to achieve personal recognition). In a context of economic change and uncertainty, which will inevitably require firms to reinforce flexibility, organise themselves in networks and be more responsive to human needs, SMEs have many advantages. The 21st century will no doubt also be marked by the changing role of women, and the growing number of women entrepreneurs world-wide who will probably become a source of change and innovation in the business world and in society. **Thierry Gaudin [France, Business Environment and Technology]** stressed that technological changes lead to new ways of organising time and communications which may more closely resemble a woman's way of doing things.

To maintain competitiveness, SMEs need to develop tangible and intangible investment and strategic capabilities (...) requiring appropriate management to locate and develop the necessary resources.

SME competitiveness depends on a series of factors related to: *a)* tangible investment in technologically appropriate equipment; *b)* intangible investment which includes the ability to access technology and software efficiently; the ability to use information about the market and competition efficiently; staff training; quality of business organisation; R&D capacity and the ability to liaise with research institutions; *c)* strategic capabilities, *i.e.* innovation and flexibility. The competitive strategies of SMEs can vary widely depending on the type of firm, sector, location, market, regulatory framework and economic situation. Competitiveness strategies require appropriate management to locate and develop the necessary resources.

How do these factors affect women-owned SMEs?

The ability of women to start and develop their own businesses and make them competitive must be reviewed in light of these issues. Information and discussion in the various workshops indicated that there is, as of yet, no extensive knowledge about how these factors specifically affect women-owned businesses. However, the importance of management and the use of technology and networks were discussed extensively in various workshops, in particular in Workshop 1, Issues in the Business Environment and Technology and Workshop 3, Management Issues.

Women entrepreneurs appear to have specific management methods and enterprise structures

Women seem to structure their businesses differently, to establish flatter structures, use consensus building approaches to decide on policy and practice ...

Organisation and people management are key to remaining competitive. **Candida Brush [United States, Management]** indicated that women and men differ in the practices they use to achieve performance results. Women seem to structure their businesses differently, to establish flatter structures, use consensus building approaches to decide on policy and practice. There is also evidence in the United States that women are more ready to hire a diverse workforce and offer a different blend of benefits (such as job sharing, tuition reimbursement, etc.) to employees. Women entrepreneurs are good for the social well-being of society, are often considered to give higher priority to organisational and social efficiency, and emphasize internal communication more than their male counterparts. These differences, combined with women's ability to listen, their experience in striving for harmony within their environments, seem likely to help adapt well to the evolving requirements of managers in the 21st century. As **Ambassador Swanee Hunt [United States, Dialogue...]** proclaimed, "women are the glue in the fabric of society. They take their own individual experience and create something that connects to other people. The ability that most women have to question assumptions is critical for entrepreneurs, and something we ought to encourage women to do."

... but also tend to have less time to devote to their business and to be more reticent about seeking counselling and advice.

Other factors are indirectly related to management, for instance, women tend to have less time than men to devote to their business. Indeed, in OECD and non-OECD countries alike, women continue to be largely responsible for family, household and child care – which cuts into their time for doing business. Women also tend to be more reticent to take the time to seek counselling and expert advice when starting and developing their businesses, often because these services do not target women-owned SMEs, are provided in strongly "male" oriented settings, and are not adapted to the specific constraints experienced by women. Women entrepreneurs therefore tend to take an extremely pragmatic approach to time-saving technological tools.

Technology is key to competitiveness for women entrepreneurs

The new technologies compensate for size and distance and enable companies to grow and to go global ...

Technology and information systems are bringing about deep-seated changes to the way in which small firms do business, especially for women entrepreneurs. Technology increasingly plays a role in all aspects of competitiveness: products; production techniques; management methods; firm organisation; staff training; market information. **Ingrid Andersson [Sweden, Globalisation]** explained that new technologies compensate for size and distance and enable companies to grow and to go global.

... and change the way women work.

Technology especially changes the way women manage and execute their business agenda. **Andrina Lever [Canada, Globalisation]** emphasized that electronic conferencing, the Internet, electronic commerce (which can present the "business" in extremely neutral terms), electronic networking (homepages) and "home" working via Internet, are some of the key technological innovations for women entrepreneurs' competitiveness and allow them to create and continue to develop their business. Outsourcing and telecommuting could soon make going to work each day an anachronism.

Women-owned businesses, and business in general, will increasingly use Internet as a key tool for access to information and business opportunities.

By the year 2000, there will be about 200 000 000 Internet users. The Internet is increasingly used as a key tool for access to information on quality norms, legal and regulatory requirements, fiscal regulations and opportunities, and as a learning tool for best practices and as an address/contact book. **Kaori Sasaki [Japan, Business Environment and Technology]** explained that women running SMEs often have to overcome several obstacles simply to get to the starting line at the same time as male counterparts. The Internet helps women get there faster, helps them stay competitive; it also brings people with physical disabilities and people living in remote areas to the starting line faster. **Judith Maestracci [Australia, Access to Markets]** pointed out that: "there are literally thousands of women's business organisations. Currently, there seems to be no way to identify all these organisations. I believe in the Internet concept of bringing all this information together, so that women in any country can access it and find out what the models are, what is going on, where best practice activities are occurring. The network, apart from anything else, is the major issue."

Policy Implications and Good Practices

Falling telecommunication costs benefit business, but new technologies require new skills.

Telecommunication deregulation world-wide will benefit small entrepreneurs because of the relatively high cost of using Internet and telecommunications in some areas. However, learning to use the Net and electronic commerce requires great skill, adaptation, and tangible investment – again, training and education are crucial. But, as **Jocelyne Attal [United States, Business Environment and Technology]** explained, not all women have access to a personal computer in order to do business on the Net, especially in developing and transition economies. Therefore, there is need for networked computers and Internet work stations. In addition, the Net is not necessarily as conducive to business as it might be. The designers of the Internet and the service providers must make efforts to make it more user friendly and more conducive to business.

Internet provides information on international best practices ...

Sherrye Henry [United States, Business Environment and Technology], **Judith Maestracci [Australia, Management]** and **Arline Wouterz [United Kingdom, Access to Markets]** all pointed out that an international homepage with best practices, information and case studies from women-led and -owned enterprises would be extremely useful. *Les Femmes Chefs d'Entreprises Mondiales* (known as FCEM, the world association of women entrepreneurs) has a homepage and its members in 33 countries are all producing homepages interlinked with the FCEM homepage. **Etta Carignani [Italy, Dialogue...]** explained that this has permitted them to be interactive; they now have a marketing list of around 45 000 members who do business and network with each other. The FCEM linked up with NAWBO and as **Suzanne Taylor [United States, Business Environment and Technology]** indicated, members participated in trade missions to Frankfurt, Madrid, Mexico, Portugal and Tokyo.

... and electronic networks for investment and procurement.

In the United States, Acenet is an electronic network for investors to find women-owned firms in which they would like to invest, while Pronet allows women-owned firms to input data about their firms in order to procure government contracts. **Sherrye Henry [United States, Business Environment and Technology]** believed that these two new technological advances will greatly help women in the United States, and that they could be adapted in other countries.

However, women's networks need to be strengthened ...

These examples stress the links between competitiveness, use of electronic technology, and networks, both real and electronic. However, as **Lorraine Martin** emphasized, women often underestimate the value of networking. **Sonia de Avelar [Brazil, Dialogue...]** stressed that: "we need to strengthen women's associations and networks, especially in terms of connecting to local and national associations through international networks, and a north-south effort of coalition building, to push forward women's collective access to economic resources and empowerment."

... and women need to become more involved in technology design.

Although women use technology, equipment and machinery in their enterprises, there are few examples of women who invent, design, or influence the design of equipment, machinery, or technology, as **Helen Appleton [United Kingdom, Business Environment and Technology]** pointed out. Women should link up with artisans, engineers and designers so that their specific needs, perspectives and aspirations can become part of the technology design process, which currently tends to be cut off from users, especially the smaller users. **Amy Millman [United States, Business Environment and Technology]** urged women to become immersed in technology issues in order to stay competitive. Using technology is not enough, women should create technology and manage R&D teams.

Recommendations:

Conference participants formulated a number of proposals and policy recommendations aimed at promoting the competitiveness of SMEs in general, and women-owned SMEs in particular:

- ◊ *Continue to investigate the links between technology developments and the creation of business opportunities for SMEs and women-owned SMEs.*
- ◊ *Facilitate access to technology and widespread availability to entrepreneurs and especially female entrepreneurs, low-income women and single mothers. Provide access to and training on advanced technologies for SMEs. Ensure that education and training policies pay attention to technology know-how and teach access-to-information skills.*
- ◊ *Take into account the needs, wants and requirements of women entrepreneurs when formulating information technology and networking programmes.*
- ◊ *Promote alliance building and mentoring between associations, industry and government. Create, develop and expand associations and networks which assist women business owners to access technology in order to increase competitiveness.*
- ◊ *Undertake comparative research studies on management using shared practice and lessons learned, including the collection of statistics.*
- ◊ *International organisations such as the OECD should establish best practices on the use of technology to enhance competitiveness as well as disseminating best practices on guidance, support and education for women who would like to enter the market of high-technology businesses.*
- ◊ *International organisations such as the OECD should establish best practice recommendations for OECD governments and their aid agencies on specific features of women's management, how these will fit with the economic and social demands of the 21st century, and how training programmes should be adapted to the specificities of women entrepreneurs.*

E. Facilitating Access to International and Domestic Markets

Key Issues

Globalisation of economic activities (...) enables firms to enter new markets, to exploit their technological and organisational advantages, and to reduce business costs and risks.

Although the flexibility of SMEs gives them many advantages in adjusting to market evolution, market access remains problematic in a competitive environment which does not always work to their advantage. Today's globalisation of economic activities, implying an evolving pattern of cross-border enterprise activities – international investment, trade and strategic alliances, production, sourcing and marketing – enables firms to enter new markets, to exploit their technological and organisational advantages, and to reduce business costs and risks. Technological advances, the liberalisation of markets and increased mobility of production factors underlie this expansion and are partly driven by it.

Women-owned SMEs are just beginning to develop their exports....

SMEs are not usually associated with international activity but in fact, already play an important international role, directly producing about 26 per cent of OECD exports and about 35 per cent of Asia's exports. For women-owned SMEs, evidence suggests that their degree of globalisation is very limited; women are just beginning to develop their exports. However, as **David Bowie [United States, Globalisation]** emphasized, evidence in the United States shows that women-owned firms involved in exporting generally are more established, have higher sales, more employees, and are on a higher growth path than those that are not not involved in international trade.

...but new ways of doing global business could allow more SMEs – including women-owned businesses – to successfully compete internationally.

Changes in communications, technology and the organisation of production have all favoured the process of globalisation; more SMEs, including women-owned businesses, can successfully compete internationally. **Andrina Lever [Canada, Globalisation]** pointed out that SMEs are enhanced by their ability to do business internationally thanks to technology, partnerships, franchising, licensing, network marketing, foreign distributors, social alliances, joint ventures, supplying other companies and countries on a second- or third-tier basis. Niche players who demonstrate the agility, speed, imagination, customer focus and interactive marketing skills to envision and shape new markets will be rewarded. Women entrepreneurs should be well positioned here.

> "Multinationals seldom employ women at the head of their International Divisions. In addition, in France, SMEs account for only 20 per cent of international trade. It is therefore vital that women have greater access to intensive training and experience in the international field." **[Christine Chauvet, Directeur général, Centre Français du Commerce Extérieur, France]**

Migration favours the internationalisation of business ...

Andrina Lever pointed out that the globalisation process is also favoured by the world-wide phenomenon of migration waves that have amplified in the recent years. Increased migration creates ethnic communities which maintain cultural ties and personal contacts with their native countries, and ethnic groups in host countries that take up business opportunities.

... although barriers to trade and investment limit market access.

Market access remains an issue for many internationalised SMEs. These firms often face trade and investment barriers, however, particularly in developing countries. In addition to tariff barriers and customs regulations, non-tariff barriers and many other factors – local approval procedures, standards, the costs of establishing a legal entity, the difficulties of using flexible organisational arrangements (franchise alliances), cultural impediments to distribution channels or resources – may limit market access either intentionally or unintentionally. Regarding this issue, **Ingrid Andersson [Sweden, Globalisation]** pointed out that the special features of female entrepreneurs in regard to human relations and the ability to build trust, can influence their entry in foreign markets. She noted that the interaction between the characteristics of industries and those of countries/cultures may create both advantages and opportunities as well as barriers and difficulties in international trade, depending on specific circumstances.

It appears that women encounter specific barriers, due in part to persisting discriminatory practices and to their household responsibilities.

SMEs face many challenges that are even more difficult for women-owned SMEs when they export. Their major obstacles are more often linked to their size – these businesses traditionally do not grow to a size where it is cost-effective to become international – activity, and access to information. Evidence does however suggest that women encounter specific barriers, due in part to persisting discriminatory practices and to their household responsibilities. **Angeline Low [Australia, Globalisation]** pointed out that "although women believe they have particular skills that advantage them in dealing with people in business and government in other countries", "they must balance their entrepreneurial activities with their family responsibilities and deal with cultural differences in attitudes, which can present special problems for them in many countries." In addition, conference participants stressed that women entrepreneurs face special problems in finding financing for their projects because many women-owned businesses go into or start in the service or retail sector, which financial institutions are reluctant to finance. As suggested by **[Uma Reddy, India, Dialogue...],** "A credit guarantee scheme by importers implemented through the international banks, acting as guarantors to the national local banks for the credit needs of women going into exports" would facilitate borrowing by women wishing to export.

"As we look at the growth of women-owned organisations, we should also keep an eye on those hundreds of thousands of women world-wide who are still trying to break into the market of entrepreneurship itself." **[Nancy Carter, United States, Access to Markets]**

Women's networks open up important avenues to international trade.

Male entrepreneurs and executives have long had access to associations, networks and clubs (business, social or sports clubs) where they discuss business and negotiate deals. Women entrepreneurs are generally excluded from these "old boys' networks" and informal business networks, and increasingly set up their own networks: *Les Femmes Chefs d'Entreprises Mondiales* (FCEM), headquartered in the Netherlands, and the International Federation of Women Entrepreneurs (IFWE), headquartered in New Delhi, are the two major international groups. Thousands of primarily local groups of women entrepreneurs exist around the world. These networks, through personal contacts and the sharing of know-how, open up important avenues for women entrepreneurs.

> "Without co-operation there can be no development. Entrepreneurship by women is critical to economic productivity. The association *Les Femmes Chefs d'Entreprises Mondiales* (FCEM), is an organisation dedicated to the promotion of women in business ownership and influence. Our energies are directed towards networking, training, mentoring and education. Our objectives include: *i)* strongly encouraging training for women in entrepreneurship in all OECD countries, by providing the necessary information about access to entrepreneurship and self-employment; *ii)* stimulating access to credit and financing during the start-up and early stage of female enteprises; *iii)* improving data and statistics on women business owners, especially in SMEs." **[Joke van den Boer, President, *Les Femmes Chefs d'Entreprises Mondiales* (FCEM), Netherlands]**
>
> "In India, the traditional view of women has tended to be rather conservative. Today, the Indian woman entrepreneur is making a place for herself in the emerging industrial society. The International Federation of Women Entrepreneurs is a recent organisation created in 1993 which has made its existence known all over the world." **[Rajni Aggarwal, Secretary-General, International Federation of Women Entrepreneurs (IFWE), India]**

Governments are starting to encourage access of women-owned businesses to public procurement.

Public sector procurement represents an important market for both large firms and SMEs, but because of discrimination and more subtle barriers, women-owned enterprises are only beginning to access these opportunities. In certain countries, governments actively encourage the access of women-owned businesses to public procurement. While the United States has had at least nominal policies in support of increased federal procurement for women entrepreneurs for two decades, women-owned businesses still receive only a small percentage of all federal procurements. Despite this, significant progress has been made in the last several years as a result of concerted effort by federal policy makers. It appears that some of the key areas they are acting upon include: emphasizing the participation of women-owned businesses, standardizing procedures and promoting training.

However, women can find themselves excluded from corporate business opportunities as they are not part of "boys' clubs".

The problem of access is also apparent in winning outsourcing contracts by large companies. Large corporations may have aggressive business partner, supplier and outsourcing programmes, but because women entrepreneurs do not tend to be part of influential – primarily male – business networks, they often find themselves excluded from corporate business opportunities. This may also be true for second-tier and third-tier subcontracting programmes developed by large multinational corporations which have encouraged SME suppliers of goods and services to begin exporting (for example, if a prime contractor has sold equipment to a firm in country X, a second-tier exporter might provide replacement parts and the third-tier exporter might provide training in use of the equipment).

Policy Implications and Good Practices

Policies need to ensure that women have educational opportunities, information and access to markets.

In many countries, policy and decision makers are increasingly concerned with helping women in the area of exports, and also with improving their access to domestic markets at national and local levels. The implication is to ensure that women have educational opportunities, information and access to markets. A number of successful initiatives and government or public programmes were presented during the conference.

Some countries have developed gender-neutral and gender-specific programmes exist to assist companies to export ...

For instance, in the United States, **David Bowie [United States, Access to Markets]** explained that there are many gender-neutral programmes to assist companies to export or to increase their exports. The cornerstone of the women-in-business initiative has been a series of "Women in Trade Business Development" missions which provide excellent opportunities for participants to establish business contacts overseas. The "Women in Trade" initiative also includes outreach to women's business organisations to inform them of export assistance programmes such as the US Trade Information Center and the Trade Base Programme (toll-free telephone number, Internet and broadcast fax information system to assist potential exporters or importers with a wide range of counselling on federal export assistance programmes).

... and promote awareness of and access to existing programmes.

The Australian authorities are also concerned with ensuring that women business owners are aware of and have access to the full range of women in export programmes, according to **Angeline Low [Australia, Globalisation],** who described how AUSTRADE, the government agency to promote trade activities, encourages business women to export; it identifies business women active in exports who can act as role models for other women in business and tries to help them network with women in other countries to create a greater export focus. One of the main problems is that women tend to underestimate government, assistance and trade services.

Electronic networks aim to broaden participation in international trade.

International organisations also encourage such trends through electronic networks. UNCTAD's electronic network, Trade Points, was set up with the aim of broadening participation in international trade. **Christine Cattell [UNCTAD, Globalisation]** explained that Trade Points are basically trade facilitation centres where traders can obtain information on how to carry out international transactions. Trade Points are also a gateway to international networking and, when Trade Points are set up in various countries, women's organisations and business associations seek to participate in the programme.

Improving access to markets for women-owned businesses also includes ensuring access to public procurement.

Anthony DeLuca [United States, Access to Markets] described a creative programme established by the US Air Force to ensure access to public procurement for women entrepreneurs. Mr. DeLuca explained that to help ensure that women-owned businesses had information on pending procurements, the US Air Force introduced the "rule of one", *i.e.* that each of their buying activities would solicit at least one women-owned business for every competitive acquisition above the Simplified Acquisition Threshold (SAT). Further, large businesses were given listings of women-owned businesses for subcontracting opportunities. In addition to aid in the standardization process, women-owned businesses were given access to USAF activities through its electronic bulletin board, available on the Internet. Training is another important area in need of continued attention, both for USAF buyers and women business owners. Women business owners were trained by the Air Force Business Education Team in how to do business, and USAF buyers received training to increase sensitivity when dealing with women-owned businesses. These efforts at improving the access of women to USAF procurement resulted in a 25 per cent growth over a two-year period.

Recommendations:

Conference participants generally agreed that access to financing and guarantees for exports were considered the cornerstone in helping women to develop their international activities. However, a number of other related specific proposals/recommendations were formulated to encourage women to export and to improve access to their domestic markets. In particular:

◊ *Provide comprehensive education/training programmes.*

◊ *Provide mentoring programmes. An "international mentor-protégé programme" should be adopted to help women-owned businesses by sharing and building the necessary foundations for growth and by providing strategic counselling.*

◊ *Recommend that governments reduce red-tape and administrative burdens for women-owned SMEs which are willing to export or access new markets.*

◊ *Increase the participation of women in public procurement.*

◊ *Encourage governments to promote women's trade missions and their participation in all official trade missions.*

◊ *Encourage the formation and development of associations and networks of women entrepreneurs aimed at creating an international trade infrastructure through educational programmes, improved access to credit, identifying target markets, developing on-line databases that identify buyer-seller opportunities, licensing export automation systems, and identifying reliable joint venture partners.*

◊ *Encourage strategic alliances/joint ventures between migrant women working overseas and women in their native countries to foster greater business opportunities.*

◊ *Use electronic commerce and create a global market-place on a technological network for women-owned businesses. Open communication and information exchanges should be carried out in a practical and efficient manner.*

◊ *Recognise the need for organisations to be able to facilitate international exchanges of experts between countries, and in particular for developing and transition economies, on the issue of women's entrepreneurship.*

◊ *Improve information about current assistance programmes to enable small firms to access markets, and increase the utilisation rate of these assistance programmes by illustrating and communicating more effectively how and why the services will benefit women.*

◊ *International organisations such as the OECD should encourage governments to share information on the needs of women business owners in accessing markets and to promote the diffusion of best practices for market access for women-owned businesses in OECD, transition and developing economies.*

F. Transforming the Regulatory and Institutional Environment

Key Issues

Women need to lobby in order to obtain the types of policy they require for their small businesses.

The conference focused on the activities of women entrepreneurs, but it also clearly showed that women want to be able to influence public policy so that it incorporates *their* vision of how the economy should be structured, the main future social and economic orientations, and their specific needs so that they can contribute to this vision. Most workshops therefore discussed issues related to institutional and regulatory frameworks conducive to a favourable business environment for women entrepreneurs.

Small firms spend a disproportionately high amount of their time in dealing with administrative and regulatory matters.

SMEs frequently complain about regulations and administrative burdens on their operations. **Jane Applegate [United States, Management]** stressed that these are among the biggest challenges for small firms that have to devote a disproportionately high amount of their time to dealing with administrative and regulatory matters, which is very costly. Larger firms can spread this burden over a much higher volume of output and people, which gives them a comparative advantage. This contrasts with government policy of ensuring "a level playing field" for all firms, regardless of size. Many countries have taken innovative approaches to streamlining redundant regulations, but the regulatory process itself has often failed to keep pace with the evolution of economies and technology.

For women, regulatory and administrative burdens appear to be even heavier because they tend to set up small and micro enterprises and because they may have less experience in acquiring information about relevant administrative regulations and in carrying out many administrative tasks. **Naoko Bano [Japan, Management]** illustrated the credibility issue for women entrepreneurs by describing the difficulties she met when trying to acquire the information she needed for product approval from her local government office.

The legal status of women in business needs to be clarified.

In addition to these burdens which affect all SMEs, women entrepreneurs occasionally face other difficulties because of their legal status. Indeed, the choice of a legal structure for a business is a vital factor which may affect levels of taxation, profitability, social security and retirement benefits. Female co-entrepreneurs are often dependent on their partners for these social benefits, and in many countries the fiscal system is ill-adapted to their status as women co-entrepreneurs. **Maria Teresa d'Avila [Portugal, Financing]** pointed out that recourse institutions are unable to settle such disputes because the law does not explicitly address this issue. Therefore, legal recognition of women's role and rights in relation to other family members in a family-owned business is a crucial issue.

A flexible and responsive regulatory framework is a prerequisite.

Participants pointed to the need for a more flexible and responsive regulatory framework for very small firms or micro enterprises, typical of the firms women own. More flexible and responsive regulatory frameworks, and lower non-wage labour costs would also help women entrepreneurs in the start-up phase, as **Florence Lardenchet [France, Dialogue...]** pointed out. In transition economies, regulatory rigidity and the lack of an appropriate business framework are serious obstacles for women entrepreneurs and very small firms. According to **Josianne Capt [ILO, Dialogue...],** in developing countries, most micro enterprises headed by women are unlikely to grow into larger firms, many of these enterprises are in the informal sector and disregard most of the basic international labour standards (notwithstanding the fact that in many developing economies women cannot own land and their access to production means is limited). The issue is whether or not these micro enterprises can be brought to the level where they can become formal enterprises which respect labour standards and provide social security benefits.

More effective support structures are needed for women-owned businesses. **Lois Stevenson [Canada, Dialogue...]** pointed out that women have to a large extent responded to this need themselves by forming associations to access markets more effectively and to transform the regulatory and institutional environment. These associations also provide support, mentoring and training, and catalyse opportunities for women entrepreneurs to begin building networks with government and corporate procurement officials.

"In many countries, women entrepreneurs do not use the existing networks or the existing resources. How governments can more effectively work with associations of women entrepreneurs and others, so that they are not re-inventing the wheel, but basically buying and tying pre-existing resources." **[Virginia Littlejohn, United States, Globalisation]**

Women recognise the need to develop, command and lobby for policies and initiatives. Their associations and networks are improving their own knowledge about advocacy processes and methods for influencing public policy legislators, both in terms of executive and legislative advocacy at the local, state and national levels, and the tenets of effective advocacy at the multilateral and international levels.

"If women are among the drivers of the economy – paying taxes, using credit, creating jobs and money – then the government will hear our voice. How do we legitimize the leadership style that women may bring, and how do we get to the tables where some of the real decision-makers are – including Central bankers, CEOs or Politicians ? Are the Women SMEs at the table of the World Economic Forum ? at the International Chamber of Commerce ? At the Multinational's boards ? because much of the decision making power reside there. Are there other organisation that we can create that may help set the world agenda ? Women leaders often symbolize change in a way current people in power never do." **[Laura Liswood, United States, First Plenary Session]**

Policy Implications and Good Practices

Through associations, women are now starting to influence the decision-making process.

Women have achieved major steps in the areas discussed above in some countries. In the United States, the National Women's Business Council, as presented by **Amy Millman [United States, Business Environment and Technology]**, was set up to create an opportunity to influence the decision-making process for groups like the National Association of Women Business Owners (NAWBO) and 700 other women's business organisations. It ensures that government policy reflects their priorities and that other institutions in the private sector are cognisant of their contributions. A forum brought together NAWBO and other women associations with banking and other financial organisations and educational institutions to create a women-led agenda, designed by women illustrating what they believe the economy will look like in the next decade, what they wish it to look like, and the supporting infrastructure necessary to get there.

The leaders of this initiative work closely with the US Interagency Committee on Women Business Enterprise, which comprises top officials at major federal government departments and agencies, to discuss how to translate the needs of American women into action, policies and programmes. The next steps are to develop local networks that can disseminate information and teach women about the opportunities that the federal government makes available to them and about information conducive to the growth of their businesses. In the next two years, they will work on issues related to growing women's enterprises.

Business training programmes for women are gaining ground ...

Susan Davis [United States, Dialogue...] reported that the Center for International Private Enterprise (CIPE), part of the US Chamber of Commerce, has also developed business association training programmes for women. In the last ten years, the overall participation of women in CIPE business association training has jumped from 7 per cent in 1986 to 32 per cent in 1995. Beyond this type of training, CIPE also provides direct grants to organisations to assist them in their development. The CIPE aim is to promote a very even economic playing field for women entrepreneurs, effectively increasing the strength of women's associations. Indeed, for CIPE, business women can and must shape public policy through business organisations and public policy advocacy, capacity building, governance, utilisation of the media, financial management, etc.

... and women are increasingly involved in top-level debates.

The Australian Council of Business Women (ACOB) was set up in November 1995 to ensure that the opinions, ideas and missions of Australia's businesswomen are included in top level debates, and to influence the creation of wealth and employment. The Council addresses how to develop, command and lobby on policies and initiatives, develops and disseminates relevant information and research, and promotes the achievements of women in business. It also addresses how to participate in international networks **[Angeline Low, Australia, Globalisation]**.

Governments in some countries have implemented initiatives to support women entrepreneurs.

Uma Reddy [India, Dialogue...] indicated that the Indian government has implemented a number of initiatives to support women entrepreneurs, including the use of fiscal policies. Women entrepreneurs have also organised themselves into associations to help one another. AWAKE, the Association of Women Entrepreneurs of Karnataka, is based on the voluntary efforts of successful women entrepreneurs. It provides support in training, business and counselling, and offers a business incubator. AWAKE has its own research and resource centre, and a management development programme for growth for women entrepreneurs. It is also affiliated with the Women's World Banking, has had successful results, and has been asked by the government to undertake a rural industrial programme.

Recommendations:

Participants in the conference workshops made many policy recommendations aimed at transforming the regulatory, institutional and business environment. They focused in particular on the role of women in influencing general public policy and framework conditions, and considered how to help SMEs in general, and women-owned businesses in particular:

- ◊ *The culture of entrepreneurship should be further developed, to foster the ability of individuals to seize new business opportunities, risk-taking and innovation.*
- ◊ *Governments should regularly review legislation in order to ensure that policies are not biased towards large enterprises which can restrict competition by eliminating small businesses and self-employed competitors. The role of fiscal policy in helping women entrepreneurs should be taken into account.*
- ◊ *Encourage and create advocacy roles for organisations to develop legislation, reduce regulations affecting SMEs and create procurement opportunities in the private and public sectors.*
- ◊ *Women business owners should become involved in reducing the "red tape burden" in the private sector, as administrations in many countries are already working with the private sector to identify and reduce the regulatory burden.*
- ◊ *Women entrepreneurs, particularly those in small or home-based businesses who can feel isolated, should know more about and have better access to existing networks and resources for training and know-how, technology diffusion, developing contacts that can facilitate access to markets, etc.*
- ◊ *Women's associations and networks should be strengthened, especially in terms of connecting local and national associations through international networks, to push women's collective access to economic resources and empowerment.*
- ◊ *International organisations such as the OECD should encourage Member governments to include the women business owners' dimension in the development of government policy for SMEs, including policies in development co-operation efforts.*
- ◊ *International organisations such as the OECD should help mainstream women's issues in non-member countries, developing and transition economies, and make them central to development agendas. It should also promote the policy dialogue on these issues between recipient and Member countries.*

Conclusions

Together with men, women should invest in human resources, technology, and knowledge creation and the implementation of the necessary new structures for broad-based economic development.

The conference opened a real dialogue among policy makers, entrepreneurs, and institutions about women and the creation of businesses and wealth. Through this conference the OECD is contributing to the realisation of the potential of female entrepreneurship. Women entrepreneurs strongly stated that they do not depend on governments alone to attain this goal. Together with men, they should invest in human resources, technology, and knowledge creation and the implementation of the necessary new structures for broad-based economic development. Connections must be created among women and men in enterprises, across countries, and between developed and developing countries.

> "Today, we are talking about women and the creation of businesses and wealth, touching on the very sense of economic policy at the OECD. That historical change is a reflection of what has gone on in terms of our understanding of economic societies, of the important role that women have to play in maximising the talents and opportunities for our economies in remaining competitive in the future." **[Chris Brooks, OECD (TDS), Closing Plenary Session]**

There is a need for a forward-looking vision and for building an agenda for change ...

Conference participants agreed upon the necessity of a forward-looking vision and endorsed the effort to build an agenda for change. "We have to look at the year 2000 as our initial step on what we want to move ahead with the policy recommendations that have come out of this conference" **[Ronnie Feit, United States, First Plenary Session]**.

... as well as an action plan and accountability.

Some recommendations need to be implemented at the government level and at the multilateral level, but the private sector – private banks, corporations, and associations of women entrepreneurs – also have a role to play in implementing these recommendations. "We need not only a common vision or a strategy for implementation, but also real determination, an action plan, a timetable for when we want to do things, and an accountability for who is following-up on things" **[Virginia Littlejohn, United States, Closing Plenary Session]**.

> "If women's businesses keep growing as they have been, we can imagine owning at least half of the assets in the world. The Chinese have a proverb that says women hold up half the sky. We can hold up our half better if we own the ground we stand on." **[Ronnie Feit, United States, First Plenary Session]**

Recommendations of the Conference

The following represent the consolidated recommendations that emerged from the conference:

◊ ***Improve understanding of the contributions of women entrepreneurs in OECD, transition and developing countries to economic growth, international trade and job creation through strengthened data and statistics about women-owned SMEs. Also, increase statistical harmonization across countries in order to facilitate comparative analysis.***

◊ ***Improve understanding of the impediments faced by women-owned businesses, and how they have successfully been addressed by business and policy makers.***

◊ ***Promote government policies to provide a level playing field and improve competitiveness for women-owned businesses, e.g. by taking steps to remove distortive legislative and regulatory barriers.***

◊ ***Remove excessive barriers to obtaining capital faced by women entrepreneurs, and develop financial instruments that address legitimate capital needs at all stages along the business continuum (micro, small, medium and large enterprises). Include a special focus on the financial needs of businesses in the service sector. Research should compare experience among countries in OECD, developing and transition economies, and analyse the nature and value of "intellectual capital" so as to provide financial institutions with new elements to consider in evaluating credit risks.***

◊ ***Welcome private initiatives for an international consortium of banks, aiming at raising financial sector competence and awareness regarding financing of women entrepreneurs.***

◊ ***Encourage aid agencies and multilateral regional development banks to develop programmes permitting equal access to capital for women entrepreneurs; equal opportunity should also apply to programmes relating to data and statistics, management and technical assistance, association building, technology utilisation, access to markets and international trade.***

◊ ***Develop appropriate benchmarking practices that recognise and address the unique contributions and needs of women-owned businesses world-wide. Benchmarking practices relating to financing could include the conditions for business angels, equity or quasi-equity capital formation, and creative guarantee programmes with appropriate training.***

◊ ***Identify global "best practices" concerning women entrepreneurs, and syndicate these practices broadly via associations and NGOs, the Internet, governmental and multilateral stakeholders. The OECD should assist in the gathering and dissemination of information on global best practices.***

◊ ***Encourage the educational system at all levels to provide entrepreneurial training and general economic literacy to both sexes throughout the educational process, including "best practices" and case studies of success.***

◊ ***Encourage the development in OECD, developing and transition economies of affordable and accessible educational and training initiatives to assist women entrepreneurs in business management skills, access to markets and technological know-how which can enable them rapidly to advance their businesses. Training and education ought to be flexible enough to meet unique local needs and address training needs along the business continuum.***

◊ ***Encourage the formation and development of associations and networks of women entrepreneurs for purposes of training and know-how, technology diffusion, development of contacts that can facilitate access to markets, etc. National associations should be encouraged to promote increased visibility for women entrepreneurs, and facilitate the appointment of women as advisers to government and to key decision-making roles. In addition, encourage women entrepreneurs to participate in mainstream business organisations such as Chambers of Commerce.***

Recommendations of the Conference *(cont'd)*

◊ ***Develop mechanisms to facilitate access to markets, including corporate and federal procurement.***

◊ ***Develop initiatives to facilitate access to international markets, including trade missions, international linkages between associations of women entrepreneurs, electronic commerce, etc. Governments should facilitate the availability of finance for women entrepreneurs involved in international trade (export-import finance, including export guarantee insurance).***

◊ ***Develop a research agenda on women entrepreneurs, and encourage experts, governments, development agencies, multilateral organisations, and associations to begin "bridging the gap". The OECD should continue to play a significant role as a catalyst for future action and in order to ensure that this agenda is effectively implemented.***

◊ ***Reconvene a follow-up Conference on Women Entrepreneurs in SMEs at the OECD in the year 2000 to assess the progress of conference recommendations from 1997, to analyse the new global business environment, and to make recommendations for the next five years.***

Notes

1. Quoted from Ms. Shigeko Mitsuhashi's presentation to Workshop 6 on Access to Markets. Initially extracted from a handbook for women setting up businesses, Nikkei Research Institute of Industry and Markets.

2. Intergovernmental OECD Conference on "Education and the Economy in a Changing Society", 1988, and the OECD-CERI (1989) report, *Towards an "Enterprising" Culture – A Challenge for Education and Training*, which discusses many fundamental issues relating to the possibilities and obstacles involved in introducing an entrepreneurial spirit into education and training.

3. The Centre for Educational Research and Innovation of the OECD has carried out a number of activities on lifelong learning and human capital investment. See also OECD (1998), *Human Capital Investment.: An International Comparison.*

Bibliography

BARRETT, Mary (1997), Regional Report on the Asia-Pacific Area, prepared as unpublished background material for the OECD Conference.

BRUSH, Candida (1990), "Women and Enterprise Creation", in Sara K. Gould and Julia Parzen (eds.), *Enterprising Women: Local Initiatives for Job Creation,* OECD/LEED, p. 37.

CATTAN, Claude (1997), Regional Report on Italy and Southern Europe, prepared as unpublished background material for the OECD Conference.

DUCHENEAUT, Bertrand (1997), "Women Entrepreneurs in SMEs", Background Report for the OECD Conference.

HOLMQUIST, Karin (1996), "Female Enterprises: Current Situation and Development Trends" in *Nordic Labour Journal 2/1996.*

MARLEAU, Martine (1994), "L'Entrepreneurship au Féminin", rapport de la Fédération des Entreprises Indépendantes, December.

NFWBO (1997), *1996 Facts on Women-owned Businesses: State Trends*, The National Foundation for Women Business Owners, Silver Spring, MD, United States.

OECD (1994), *Women and Structural Change: New Perspectives*, Paris.

OECD (1996*a*), *Technology, Productivity and Job Creation*, Vol. I, Highlights, Paris.

OECD (1996*b*), *SMEs: Employment, Innovation and Growth, The Washington Workshop*, Paris.

OECD (1997*a*), *Labour Force Statistics*, Paris.

OECD (1997*b*), *Globalisation and Small and Medium Enterprises* (*SMEs*), Vol. 2, Country Studies, Paris.

RUNEBORG, Silvia (1997), Regional Report on Sweden and Northern Europe, prepared as unpublished background material for the OECD Conference.

SKINNER, Jane and Jane SLOWEY (1996), Report for the OECD LEED Programme, October.

SMALL BUSINESS ADMINISTRATION (1997), *Statistics on Women Business Ownership.*

STARCHER, Diane Chamberlain (1996), "Women Entrepreneurs, Catalysts For Transformation", *European Baha'í Business Forum.*

WEEKS, Julie (1997), Regional Report on North America and Selected Latin American countries, prepared as unpublished background material for the OECD Conference.

WEISS, Chris (1990), "The Role of Intermediaries in Strengthening Women's Self-employment Activities" in *Enterprising Women*, OECD/LEED, Paris.

Chapitre 1

Les Femmes Entrepreneurs à la Tête de PME
SYNTHÈSE

Lors de la session d'ouverture de la conférence sur les "Femmes Entrepreneurs à la tête de PME : une nouvelle force pour l'innovation et la création d'emplois", le Secrétariat de l'OCDE a présenté un document de discussion qui exposait l'ensemble des questions destinées à être examinées au cours de la conférence. Ce chapitre de synthèse fait suite à ce document de discussion et s'articule autour des mêmes thèmes. Tirant parti d'un ensemble très riche d'informations fournies par la conférence, il s'attache à mettre en lumière un certain nombre de problèmes cruciaux et des solutions possibles. Ce chapitre traduit les faits, les idées et les recommandations mis en avant par cette conférence et souligne la nécessité de disposer de données quantitatives et qualitatives comparables au plan international.

Introduction

La population active mondiale est pour moitié composée de femmes.

La population active mondiale est composée pour moitié de femmes. Ce chiffre qui semble être une évidence, présente en réalité des dimensions nouvelles. Au moment où les pays industrialisés s'efforcent de retrouver le chemin de la croissance économique et de créer des emplois, la gestion des ressources humaines, la créativité et l'innovation sont les pierres angulaires d'une amélioration de la situation économique et sociale.

Les PME sont une importante source d'emplois, d'innovation et de développement économique.

Les pays Membres et non membres de l'OCDE traversent une période de profonds changements et se heurtent de plus en plus aux défis de la mondialisation des marchés, de la concurrence accrue et de l'interdépendance grandissante des économies. Parallèlement, on observe dans la plupart des pays de l'OCDE que les Petites et Moyennes Entreprises (PME) sont une source importante d'emplois, d'innovation et de développement économique. Or, ce potentiel est dans l'ensemble très peu exploité et les imperfections du marché, les dispositifs mal adaptés et les incohérences institutionnelles qui jouent à l'encontre des PME sont à l'origine de nombreuses difficultés stratégiques (OCDE, 1996*a*).

Selon certaines sources, les PME appartenant à des femmes enregistrent un taux de croissance plus rapide que l'économie dans son ensemble dans plusieurs pays de l'OCDE ; toutefois, la disparition d'un certain nombre d'obstacles permettrait à leur potentiel d'être pleinement exploité.

Tel est surtout le cas des PME dirigées par des femmes qui, selon certaines sources, enregistrent un taux de croissance plus élevé que l'économie dans son ensemble dans plusieurs pays de l'OCDE. Les femmes chefs d'entreprise sont en train de devenir une force très importante dans les pays de l'OCDE, mais leur contribution deviendrait encore plus importante si un certain nombre de restrictions disparaissaient afin que leur potentiel soit pleinement exploité. Il est indispensable d'analyser leurs contributions et leurs besoins et d'appliquer les réformes structurelles de l'ampleur voulue, pour trois raisons :

◊ ***Economiques*** : Les femmes qui dirigent une PME créent de l'emploi pour elles-mêmes et/ou pour d'autres. De nombreuses sociétés investissent à long terme dans leur personnel féminin mais, en vertu du phénomène du "plafond invisible", ne font rien pour promouvoir ces femmes aux postes de direction les plus élevés. Donner à ces femmes l'occasion d'essaimer et de créer leur propre entreprise est un moyen de capitaliser pleinement sur les compétences et la formation qu'elles ont acquises. Créer son propre emploi peut également permettre d'échapper au chômage.

◊ ***Sociales*** : La création d'entreprise représente non seulement une alternative au chômage, mais permet en outre à une femme de concilier travail et responsabilités familiales, ce qui contribue au bien-être des familles et à la cohésion sociale. Cela étant dit, les femmes entrepreneurs peuvent rencontrer des difficultés en tentant d'équilibrer leur carrière avec leurs responsabilités familiales.

◊ ***Politiques*** : En aidant les PME appartenant à des femmes, on contribuera à réduire l'inégalité entre les hommes et les femmes, on accroîtra l'autonomie féminine et on permettra aux femmes de jouer un rôle plus actif et représentatif dans la vie économique et politique nationale.

Favoriser l'activité des femmes chefs d'entreprise consolide l'économie et peut être une source d'innovation politique, économique et sociale. Souvent, par rapport à leurs homologues masculins, les femmes qui possèdent une entreprise ont une autre vision de l'organisation et de la gestion parce qu'elles ont un autre vécu et une autre approche des choses. Ces qualités constituent une vraie source d'innovation potentielle sur le plan du style de gestion, de la structure de l'entreprise, des services rendus à la collectivité, de l'emploi des technologies, etc. En présentant ces recommandations aux responsables des politiques et des décisions, l'OCDE se propose d'aider les pays Membres et non membres à concrétiser le potentiel d'innovation et de création d'emplois des PME appartenant à des femmes.

"Améliorer les conditions de création et de croissance des PME dirigées par des femmes est un enjeu de taille pour les décideurs politiques. Les gouvernements n'ont pas le pouvoir de faire que les entreprises se développent et emploient du personnel. Ce qu'ils peuvent par contre faire, c'est mettre en place les conditions et les incitations voulues pour que les entreprises et les individus s'en occupent." **[Thomas Andersson, OCDE (DSTI), première séance plénière]**

Ce chapitre de synthèse de la conférence sur "Les femmes entrepreneurs à la tête de petites et moyennes entreprises : une nouvelle force pour l'innovation et la création d'emplois" se présente par thème, comme suit :

◊ Améliorer la connaissance de l'entrepreneuriat féminin.
◊ Promouvoir dans le système éducatif l'esprit d'entreprise et la formation d'entrepreneurs.
◊ Eliminer les obstacles financiers et favoriser la mise en place d'instruments financiers nouveaux, dans le but de faciliter la création et le développement des entreprises appartenant à des femmes.
◊ Promouvoir la compétitivité des PME dirigées par des femmes.
◊ Faciliter l'accès aux marchés internationaux et intérieurs.
◊ Transformer l'environnement réglementaire et institutionnel.

A. Améliorer la Connaissance de l'Entrepreneuriat Féminin

Principaux problèmes

En dépit de la richesse et de la variété d'informations collectées au cours de la conférence, il est apparu très clairement qu'il existe très peu de données et de statistiques comparables sur le plan international concernant les femmes entrepreneurs

Difficultés liées aux données et statistiques

On manque de recherches et de données statistiques sur l'entrepreneuriat féminin, ce qui limite l'analyse et oblige à travailler sur des estimations et des hypothèses *a priori.* Il est capital de favoriser des recherches permettant de mieux informer les différents acteurs et les décideurs sur un phénomène social et économique qui recèle un réel potentiel d'innovation et de création d'emplois au XXIème siècle. **[Bertrand Duchéneaut, France, Première séance plénière]**

Collecter des informations et des statistiques sur les femmes chefs d'entreprise devrait faire partie intégrante des travaux de collecte de données sur les PME.

De l'avis général, de meilleures statistiques sont indispensables si l'on veut mieux comprendre la contribution qu'apportent les femmes chefs d'entreprise à la croissance économique, au commerce international et à la création d'emplois dans tous les pays. Cette collecte devrait faire partie intégrante des travaux ordinaires de collecte de données sur les PME.

Les données sont souvent incomplètes ...

En effet, les statistiques officielles ne permettent dans aucun pays de connaître exactement le nombre de femmes chefs d'entreprise car les chiffres qui sont fondés sur des analyses d'échantillon ou des sondages sont souvent incomplets. De plus, la PME n'est pas officiellement définie partout de la même manière en l'absence de seuils nettement fixés pour l'analyse. Les entreprises peuvent être classées parmi les PME en raison de leur chiffre d'affaires, du nombre de leurs salariés et/ou de leur statut juridique. Il est encore plus problématique d'obtenir des statistiques concernant les femmes car les enquêtes ne sont pas ventilées selon le sexe. Des problèmes de définition compliquent encore la collecte des données : on peut entendre par "femme entrepreneur" aussi bien une femme ayant créé une entreprise où elle travaille seule, que quelqu'un qui est le principal responsable d'une entreprise familiale ou d'une petite société ou encore une femme qui possède des participations dans une entreprise cotée en bourse qu'elle gère.

... et il est difficile, en partie pour des raisons historiques ou au nom de la défense des libertés publiques, d'obtenir des statistiques nationales, ventilées selon le sexe, sur le revenu et la fortune.

Cette situation est en partie due à des raisons historiques ou aux lois sur les libertés publiques qui interdisent aux systèmes nationaux de statistiques de collecter et publier certaines informations. Des statistiques détaillées, ventilées selon le sexe, sur le revenu et la fortune sont, en conséquence, particulièrement difficiles à se procurer dans certains pays. L'organisation d'une collecte internationale de données à des fins de comparaison pose encore plus de problèmes d'accès et de cohérence.

Pour que le rôle économique et social des femmes entrepreneurs et de leurs entreprises soit plus apparent et mieux compris, il faut disposer de données et de statistiques.

Encore une fois, et comme cela a déjà été indiqué plus haut, l'analyse et les choix d'orientations exigent, pour être solidement étayés, des informations et des statistiques de qualité, indispensables aussi pour vérifier l'efficacité des initiatives. Faire que le rôle économique et social des femmes entrepreneurs et de leurs entreprises soit plus apparent et mieux compris – car il s'agit d'un phénomène nouveau qui pourrait ouvrir des possibilités réelles d'innovation et de création d'emplois au XXIème siècle – nécessite des données et des statistiques. Bien que les données ne suffisent pas à asseoir la crédibilité des femmes chefs d'entreprises, elles constituent la base de tout débat bien documenté.

"Le manque de crédibilité représente le plus grand obstacle à l'accès aux marchés et aux informations commerciales, aux relations d'affaires et à la position des femmes dans le milieu des entreprises ... Au sein d'un groupe de femmes chefs d'entreprise, nous avons décidé que si nous voulions qu'on nous prenne au sérieux, il nous fallait des données et des statistiques qui rendent pleinement compte de ce que nous faisions. Nous avons créé notre propre organisation pour faire connaître quelle était la contribution réelle des femmes à l'économie. En 1992, cet organisme, the National Foundation for Women Business Owners (NFWBO) a publié sa première étude, qui a été la première étude fiable sur les entreprises dirigées par des femmes aux États-Unis. Cette étude a fait litière de toute une série de mythes solidement ancrés. On connaît aujourd'hui, grâce à elle, l'apport des femmes à la santé et à la stabilité de l'économie. Moyennant quoi, le gouvernement fédéral des États-Unis s'est fixé pour objectif d'accorder aux entreprises dirigées par des femmes 5 pour cent des marchés fédéraux. Les entreprises se sont empressées de confier à des femmes des tâches de vendeurs. L'accès au crédit a aussi changé. Wells Fargo a créé un fonds doté de 10 milliards de dollars en vue d'accorder sur dix ans des prêts à des femmes chefs d'entreprise. Le rapport a eu aussi d'importantes répercussions sur l'accès des femmes à l'information, à la formation et aux réseaux. Nous pouvons également utiliser nos données comme moyen d'action pour encourager les femmes à créer une entreprise, pour changer le modèle de gestion que les femmes copient." **[Laura Henderson, États-Unis, Accès aux marchés]**

Que montrent les données disponibles ?

Les données disponibles indiquent néanmoins l'importance croissante du nombre des entreprises appartenant à des femmes dans les pays Membres de l'OCDE.

Le travail indépendant parmi les femmes donne une indication de l'entrepreneuriat féminin dans un pays (voir annexe 1). Le taux d'activité selon ce mode d'exercice a évolué de façon contrastée dans un certain nombre de pays de l'OCDE depuis 20 ans. Au cours des dix dernières années (1986-1996), l'emploi féminin non salarié paraît avoir nettement augmenté dans un certain nombre de pays. En Australie et aux États-Unis en particulier, cependant, la progression a été nette et constante depuis 20 ans. Mais ces données sur l'emploi non salarié ne couvrent qu'un aspect de la création de la gestion des entreprises dirigées par des femmes.

Aux États-Unis (...) ces dernières années, le nombre d'entreprises créées et gérées par des femmes a augmenté deux fois plus vite que le nombre de celles créées et dirigées par des hommes.

En ce qui concerne cette catégorie d'entreprises aux États-Unis, par exemple, sur lesquelles on possède les recherches les plus précises, le nombre d'entreprises créées et dirigées par des femmes a augmenté deux fois plus vite que celui des entreprises créées et dirigées par des hommes au cours des dix dernières années. Environ 8 millions d'entreprises sont la propriété de femmes qui les gèrent. D'après la National Foundation for Women Business Owners (NFWBO), un quart des emplois du secteur privé sont le fait d'entreprises dirigées par une femme.

Entreprises appartenant à des femmes aux États-Unis, 1987-96

	1987	1992	1996	Variation 1996/1987 (en pourcentage)	Evolution annuelle moyenne (en pourcentage)
Nombre d'entreprises (en millions)	4.47	6.41	7.95	+77.6	+6.6
Nombre d'emplois (en millions)	6.55	13.21	18.54	+182.7	+12.2
Chiffre d'affaires (en milliards)	681	1574	2288	+235.8	+14.5

Source : NFWBO (1997).

Il existe de nombreux indicateurs de l'importance croissante des PME appartenant à des femmes dans les différents pays.

En dépit de l'absence de données comparables sur le plan international concernant les entreprises dont les propriétaires sont des femmes, de nombreuses observations récentes montrent la place importante que tiennent ces entreprises à un moment particulier dans le temps, ou encore la croissance du nombre de femmes entrepreneurs.

Par exemple, **Shigeko Mitsuhasi [Japon, Accès au marché]** a signalé que le pourcentage des femmes entrepreneurs au Japon est passé de 2.4 pour cent en 1980 à 5.2 pour cent en 1995 (voir tableau ci après).

Femmes chefs d'entreprise au Japon, 1980-1995

	1980	1982	1984	1986	1988	1990	1992	1994	1996
Nombre de femmes	11 612	17 553	23 021	28 007	34 836	41 171	46 958	50 712	51 911
%	2.4	3.2	3.7	4.0	4.4	4.7	4.9	5.2	5.2

Source : Nikkei Research Institute of Industry and Markets[1].

En France, une étude de 1990 du Crédit d'équipement des petites et moyennes entreprises (CEPME) montre que parmi les entreprises employant entre 3 et 500 salariés (qui sont au nombre d'environ 550 000 en France), 19 pour cent de celles qui ont moins de dix ans sont dirigées par des femmes contre 15 pour cent dans le cas des entreprises de plus de dix ans. **[Henry Savajol, France, Financement].** En Australie, **Lorraine Martin [Australie, Environnement économique et Technologie]** a indiqué qu'en 1990, selon des estimations, les petites entreprises dirigées par des femmes sont à l'origine de 10 à 15 pour cent de la production nationale australienne et d'environ la moitié des emplois du secteur privé du pays. **Maria Teresa d'Avila [Portugal, Financement]** a signalé que selon l'Association des femmes entrepreneurs du Portugal (AMEP), il y a environ 380 000 femmes entrepreneurs au Portugal. Elle a également souligné qu'une étude transnationale effectuée aux États-Unis sur une trentaine de pays industrialisés et en développement avait montré que la croissance du nombre des femmes entrepreneurs avait été des plus rapides au Portugal (153 pour cent) entre 1991 et 1995.

Dans les pays non membres de l'OCDE, le développement des entreprises féminines est également patent. Le sentiment grandissant en Asie et en Amérique latine que participer à la gestion à plus haut niveau d'une entreprise familiale est légitime pour une femme alimente cette tendance. **Rajni Aggarwal [Inde, Première séance plénière]** a signalé que d'après le recensement de 1991, 10 pour cent des entreprises indiennes sont dirigées par des femmes. Elle a également précisé que d'ici à l'an 2000, ce pourcentage pourrait doubler.

Selon l'Organisation des Nations Unies, dans le milieu latino-américain des petites et très petites entreprises, les entreprises "au féminin" pourraient bien atteindre la moitié du total si on les considère globalement, secteurs formel et informel confondus **[Sonia de Avelar Brésil, Dialogue entre les pays de l'OCDE et les économies en transition et en développement : les défis de l'avenir pour les décideurs** (ci-après désigné par **Dialogue ...)]**

Il ne fait aucun doute que les femmes entrepreneurs jouent un rôle important dans d'autres pays en développement. **Leyla Khaïat [Tunisie, Gestion]** a souligné qu'en Tunisie, par exemple, les femmes concourent de plus en plus au développement du pays. Ces tendances ne sont cependant pas suffisamment attestées. On sait que les femmes participent massivement à la production agricole. Elles seraient à l'origine de 80 pour cent de la production alimentaire en Afrique, de 60 pour cent en Asie et de 40 pour cent en Amérique latine. Ce sont elles également qui commercialisent les produits alimentaires, ce qui en fait d'excellentes connaisseuses des marchés de proximité et de leur évolution.

Les femmes des économies en transition se tournent de plus en plus vers l'entrepreneuriat.

Les statistiques ventilées par sexe sont rares dans les économies en transition mais on sait que les femmes, qui ont particulièrement souffert de la privatisation des entreprises nationales et du chômage structurel qui a suivi le passage à l'économie de marché, se tournent de plus en plus vers l'entrepreneuriat, mais en partant de très bas. En Roumanie, par exemple, les femmes créent de nombreuses PME soit seules, soit en s'associant à d'autres personnes **[Raluca Bejan, Roumanie, Dialogue ...]**.

En quoi l'entrepreneuriat féminin a-t-il un caractère spécifique ?

"En quoi l'entrepreneuriat féminin est-il important ? En quoi est-il différent ? Pourquoi ces différences sont-elles importantes et que faut-il changer ? Les études de la Banque mondiale et de l'ONU, ainsi que les études empiriques faites dans les différents pays, font clairement apparaître que la ségrégation au travail existe dans presque tous les pays de l'OCDE et dans de nombreux pays peu développés. Les femmes n'ont pas les mêmes chances à l'égard du travail. Elles abordent la création d'une entreprise avec un vécu professionnel différent et se lancent par conséquent dans des projets différents. Elles ont une expérience du travail dans d'autres secteurs que leurs homologues masculins, ce qui leur complique assez la tâche lorsqu'elles veulent créer une entreprise dans un secteur qu'elles ne connaissent pas. Tenues par la société d'assumer leur rôle de femme et de mère de famille, elles se voient généralement obligées d'endosser une plus grande part de responsabilités dans la tenue du ménage, les soins à donner aux enfants et aux personnes âgées ou handicapées, autant de tâches qui peuvent être une charge pour la femme qui veut gérer harmonieusement ces responsabilités tout en s'efforçant de développer son entreprise. S'agissant de la vie périprofessionnelle, on constate que, dans certains pays, les femmes sont assez mal acceptées dans les activités qui s'organisent en marge du travail (par exemple dans les cocktails) et sont exclues des réseaux tissés à cette occasion, ce qui les prive de quantité d'informations. Ainsi, les structures sociales et les modes de socialisation des femmes influent sur le capital humain et social qui est le leur au moment où elles créent leur entreprise." **[Candida Brush, États-Unis, Management]**

En dépit d'améliorations récentes, les femmes propriétaires d'entreprise ne retiennent pas suffisamment l'attention des autorités nationales et locales, des établissements d'enseignement et du monde des affaires et de la finance qui ne prennent pas au sérieux le rôle qu'elles jouent dans l'économie. Malgré l'insuffisance de statistiques, certaines remarques d'ordre général peuvent néanmoins être faites quant aux spécificités des entreprises appartenant à des femmes et aux obstacles que rencontrent ces femmes entrepreneurs :

Dans les pays de l'OCDE, la grande majorité des entreprises dirigées par des femmes sont des entreprises de service.

◊ Dans tous les pays de l'OCDE, la grande majorité des entreprises dirigées par des femmes sont des entreprises de service. Un sondage ayant porté sur 17 000 femmes de plusieurs pays européens (Espagne et Portugal exclus) a montré qu'environ 5 millions de femmes exercent une activité non salariée. Près de 46 pour cent de ces femmes sont des détaillantes, 12 pour cent tiennent un salon de coiffure ou d'esthétique, 10 pour cent exercent une profession libérale (médecin, avocate, etc.), 9 pour cent sont des artisanes et 1 pour cent seulement travaillent dans le secteur manufacturier.

Aux États-Unis, la distribution des entreprises appartenant à des femmes montre une prépondérance des entreprises de petite taille.

◊ Aux États-Unis, environ 75 pour cent des entreprises appartenant à des femmes n'ont pas de salarié, chiffre comparable à celui de l'ensemble des entreprises américaines ; moins de 1 pour cent des entreprises féminines ont un chiffre d'affaires égal ou supérieur à 1 million de dollars des États-Unis, alors qu'environ 2 pour cent de l'ensemble des firmes américaines ont un chiffre d'affaires supérieur à 1 million de dollars. La distribution des entreprises montre une prépondérance des entreprises de petite taille ; les femmes ont des entreprises légèrement plus petites et sont légèrement plus jeunes ; parmi les entreprises qui ont des salariés, environ la moitié emploient moins de 10 personnes **[Candida Brush [États-Unis, Gestion].**

◊ Les faits montrent qu'il existe un lien étroit entre les obstacles et le type d'entreprise que les femmes choisissent de créer – les femmes qui s'orientent vers des activités jugées proches des tâches domestiques rencontrent moins de problèmes. **[Gina Zabludovsky, Mexique, Gestion]**

◊ Dans certains pays, les entreprises de femmes semblent avoir un taux de mortalité supérieur aux entreprises masculines, ce qui peut tenir surtout au type d'entreprises en question mais être aussi le reflet des obstacles que rencontrent les femmes. Dans d'autres pays, c'est l'inverse qui se produit, ce que certains ont attribué au fait que les femmes chefs d'entreprise sont moins enclines à prendre des risques. Des recherches menées aux États-Unis (NFWBO) et au Canada ont fait apparaître que 3 sur 4 des sociétés appartenant à des femmes se maintiennent en activité plus de trois ans contre 2 sur 3 des sociétés appartenant à des hommes.

L'accès au capital est un des principaux obstacles que rencontrent les femmes entrepreneurs.

◊ L'accès au capital est un des principaux obstacles que rencontrent les femmes entrepreneurs. Ce thème a fait l'objet de l'un des ateliers de la conférence (voir ci-dessous).

◊ Le défaut de visibilité et de reconnaissance de l'entrepreneuriat féminin est probablement plus grave encore dans les pays en développement où l'on n'accepte pas aisément qu'une femme ait sa propre affaire. D'où des difficultés accrues d'accès au capital et un surcroît de problèmes quand il s'agit d'acheter du matériel, de louer des locaux ou même d'embaucher du personnel.

Recommandations :

La conférence de l'OCDE a proclamé haut et fort qu'il est du devoir des responsables politiques de trouver les moyens appropriés pour favoriser la collecte de données comparables au niveau national et international et d'approfondir la connaissance de l'entrepreneuriat féminin afin d'accroître l'efficacité globale des politiques à l'égard des PME.

Quelles données et informations collecter à l'échelon national et international ?

Il est essentiel de réunir des informations de base dans tous les pays du monde et d'approfondir les études même dans les pays, comme les États-Unis, où les informations de base ont déjà été collectées

De l'avis général, des information de base doivent être réunies dans tous les pays du monde. Même aux États-Unis, où ces informations de base ont déjà été collectées, "il faut approfondir les études" comme l'a déclaré **Julie Weeks [États-Unis, Environnement économique et technologie]**. Ces données permettront de répondre à des questions telles que les suivantes :

- ◊ Combien d'entreprises ont-elles été créées par des femmes et dans quels secteurs ? Il est important de faire apparaître l'existence d'entreprises de femmes dans des secteurs d'activité traditionnellement non féminins, comme la construction et l'industrie manufacturière, et pas seulement dans le secteur des services.
- ◊ Quelles sont les entreprises qui se développent et quel est leur taux de croissance ?
- ◊ Comment peut-on décrire et classer les femmes propriétaires d'entreprise ?
- ◊ Comment la situation des entreprises qui démarrent se compare-t-elle à celle des entreprises qui ont au moins dix ans ?
- ◊ Que se passe-t-il quand une entreprise dirigée par une femme se développe et se consolide ?
- ◊ Comment le style de gestion d'une femme, qui diffère de celui d'un homme, influe-t-il sur le développement de l'entreprise et sa rentabilité ?

Il conviendrait de modifier les techniques de collecte statistique pour permettre la ventilation par sexe des données relatives aux PME opérant sur le marché international.

Virginia Littlejohn, David Bowie (États-Unis) et Angeline Low (Australie) [Mondialisation] ont estimé que les techniques actuelles de collecte statistique devaient être modifiées pour permettre la ventilation par sexe des données relatives aux PME opérant sur le marché international. L'atelier 5 sur les problèmes financiers a également recommandé de collecter, sur le financement des PME, des données ventilées par sexe. L'importance d'une évaluation des programmes qui permettrait de mieux connaître ce qui marche a également été rappelée.

Comment et par qui ?

Les gouvernements ont un rôle à jouer dans l'amélioration de la collecte des données et de l'information sur les entreprises dirigées par des femmes ainsi que dans l'approfondissement de la connaissance de ces entreprises.

Les participants ont souligné le rôle que devaient jouer les gouvernements dans l'amélioration de la collecte des données et des informations sur les entreprises dirigées par des femmes, ainsi que dans l'approfondissement de la connaissance de ces entreprises. Toutefois la nécessité d'un effort concerté des gouvernements, des associations, des entreprises privées et des organismes de recherche en vue de favoriser ce type de travail a été également soulignée à plusieurs reprises. La création d'un réseau international des gouvernements, des grandes entreprises et des entreprises dirigées par des femmes a également été suggérée (atelier 2 sur la mondialisation et le commerce international).

Les technologies peuvent faciliter les travaux de collecte et de compilation des données.

Le recours à la technologie est considéré comme un moyen propre à faciliter ces travaux. Il conviendrait d'utiliser les données existantes ou les moyens de compilation des données existants (par exemple, données de recensement et étude de la National Foundation for Women Business Owners) et d'assurer une meilleure diffusion des recherches effectuées.

Le rôle de l'OCDE

Le rôle de l'OCDE devrait être celui d'un chef de file qui centraliserait les données qui concernent les femmes et les PME et en faciliterait la collecte et la diffusion à l'échelle mondiale.

ANNEXE 1

Pourcentage de femmes parmi les employeurs et personnes travaillant à leur compte dans quelques pays de l'OCDE, 1976, 1986, 1996
(hors agriculture et travailleurs familiaux)

Pays	1976	1986	1996	Evolution annuelle moyenne 1976-86	1986-96
Australie	26.0	30.9	32.8	0.49	0.19
Belgique[1]	28.3	28.4	28.7	0.01	0.03
Canada	n.d.	n.d.	41.6	n.d.	n.d.
République tchèque	n.d.	n.d.	27.4	n.d.	n.d.
Finlande[2]	34.8	36.1	30.5	0.14	-0.56
Allemagne[3]	34.3	26.6	27.9	-0.77	0.13
Grèce[4]	18.2	16.4	19.7	-0.20	0.37
Hongrie	n.d.	n.d.	30.7	n.d.	n.d.
Italie	27.4	22.4	23.7	-0.50	0.13
Japon	33.0	35.9	32.2	0.29	-0.37
Corée	n.d.	n.d.	30.7	n.d.	n.d.
Mexique[5]	n.d.	n.d.	36.1	n.d.	n.d.
Norvège	23.9	23.4	30.3	-0.05	0.69
Pologne	n.d.	n.d.	30.7	n.d.	n.d.
Espagne	26.2	24.8	27.7	-0.14	0.29
Suède	22.6	31.2	25.9	0.86	-0.53
Turquie	n.d.	n.d	5.3	n.d.	n.d.
Royaume-Uni[6]	21.0	25.7	30.6	0.47	0.61
États-Unis[7]	26.9	33.3	39.2	0.64	0.59

Pourcentage de femmes parmi les employeurs et personnes travaillant à leur compte dans quelques pays de l'OCDE, 1976, 1986, 1996
(hors agriculture et travailleurs familiaux)

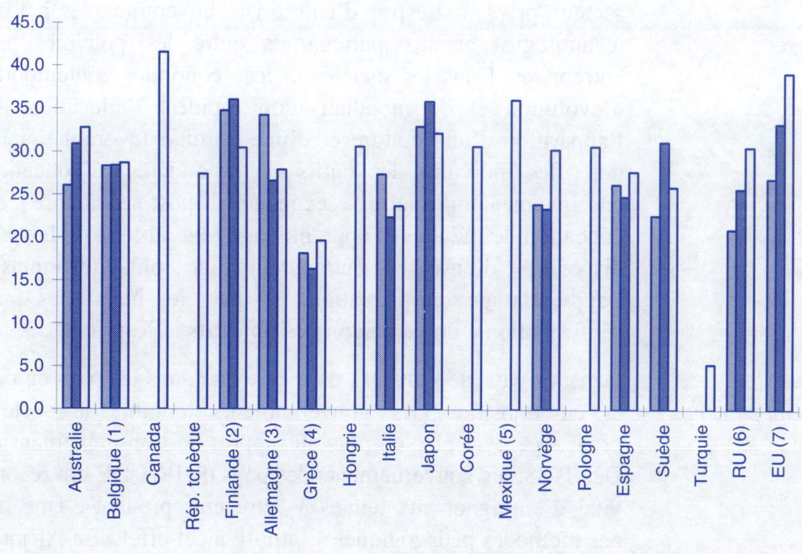

1. Le dernier chiffre se rapporte à 1992. **2.** Le premier chiffre se rapporte à 1997. **3.** A partir de 1991, les données se rapportent à l'Allemagne unifiée. Les données antérieures à 1991 se rapportent à l'Allemagne occidentale. **4.** Le premier chiffre se rapporte à 1997 et le dernier à 1995. **5.** Le dernier chiffre se rapporte à 1994. **6.** En raison de modifications de la méthode de recensement, à compter de 1990 les données ne sont pas directement comparables avec celles relatives aux années antérieures. **7.** Chiffres se rapportant à 1995.
Source : OCDE (1997*a*), *Statistiques de la population active.*

B. Promouvoir dans le Système Éducatif l'Esprit d'Entreprise et la Formation d'Entrepreneurs

Principaux problèmes

Pourquoi faut-il s'attacher spécialement à développer l'aptitude à entreprendre des fillettes et des femmes ? Y a-t-il des groupes de garçons et d'hommes qui en profitent aussi ? La société et l'économie tout entière en tirent-elles aussi avantage ? A qui incombe-t-il au premier chef de favoriser les transferts de connaissances entrepreneuriales ? Comment l'école peut-elle préparer et encourager garçons et filles à devenir entrepreneurs ? Comment et à qui inculquer des notions de base de commerce et de gestion ? Que peuvent faire les établissements d'enseignement pour utiliser, à tous les niveaux, des méthodes privilégiant l'épanouissement personnel de tous ? Telles sont quelques unes des questions dont la conférence a débattu.

> "Les stéréotypes doivent changer ; il faut favoriser l'acquisition de l'esprit d'entreprise dans toutes les écoles afin de former des individus moins passifs et moins paternalistes et leur proposer de meilleurs modèles. Il faut appliquer des méthodes de formation non classiques et bon marché à de larges couches de la population, et surtout aux femmes des régions rurales et des zones en difficulté, pour les aider à ouvrir une entreprise, à travailler chez elles ou à faire des études. Il faut utiliser partout – et pas seulement dans le système éducatif – tout ce qui peut servir à développer l'esprit d'entreprise : manuels, documentation, campagnes de motivation. Des méthodes nouvelles (magazines faciles à lire, bandes dessinées, cassettes audio et vidéo, livres, produits multimédia, Internet) peuvent encourager les femmes à vouloir créer une entreprise. Des campagnes de publicité peuvent aider les femmes à ne plus se penser comme passives et non compétitives. Les femmes modernes, actives sur le plan culturel et social, occupant des postes de cadre et de direction, peuvent contribuer à éveiller un esprit favorable à l'entreprise dans l'opinion, en particulier chez la jeune génération." [**Salo Grabinsky, Mexique, Gestion**]

L'éducation joue un rôle clé dans la transmission des compétences et des savoir-faire...

L'éducation joue un rôle clé dans le développement économique durable. C'est elle qui transmet les compétences, le savoir et les savoir-faire nécessaires pour stimuler le potentiel d'innovation des agents économiques et qui influe sur les processus de socialisation et d'adaptation aux changements. La mondialisation fait la part belle au secteur privé, à l'esprit d'entreprise, au commerce, à l'utilisation des technologies et aux partenariats entre les pouvoirs publics et les entreprises. Dans les sociétés et les économies contemporaines en voie d'évolution et de mondialisation rapides, l'éducation doit favoriser l'apparition d'une culture et d'une attitude favorables à l'entreprise et doter les individus des outils correspondants. L'éducation formelle – enseignement primaire, secondaire, post-secondaire et supérieur, éducation des adultes et apprentissage tout au long de la vie – comme les dispositifs d'enseignement/apprentissage plus informels – le tutorat, l'apprentissage en situation et par le biais des autoroutes de l'information – doivent avoir ces objectifs à l'esprit.

... dans le redressement économique et le développement...

Dans les pays de l'OCDE, de même que dans les pays en développement et en transition, les gouvernements, comme le secteur privé, se préoccupent de plus en plus du rapport de l'enseignement à l'économie. Dès 1988, les gouvernements des pays de l'OCDE ont reconnu qu'il était vital d'enseigner aux jeunes à "être entreprenant". Une transformation des méthodes pédagogiques s'impose à cet effet[2]. En Afrique, comme l'a souligné **Sarah Hage-Ali [Ghana, Mondialisation]**, la contribution des femmes entrepreneurs, importante pour le redressement et le développement économiques, et leur capacité de soutenir la concurrence sur le terrain de l'économie mondiale sont liées au développement de l'éducation des femmes.

... et l'instauration d'une culture de l'entreprise.

Des modalités appropriées d'éducation des jeunes et des adultes sont essentielles si l'on veut diffuser largement l'esprit d'entreprise. L'éducation peut créer des entrepreneurs nouveaux, capables d'acquérir des connaissance et de les utiliser de manière innovante ; elle peut encourager le travail non salarié et les créations d'entreprises et d'emplois. Savoir apprendre en trouvant son chemin à travers la multitude des sources d'information possibles est aussi un facteur essentiel de compétitivité des entreprises.

La technologie est un important moteur de la croissance économique : il faudrait encourager les jeunes filles et les femmes à s'orienter vers les sciences exactes.

Les créateurs de ces innovations technologiques qui transforment les entreprises et maintiennent leur compétitivité sont indispensables pour assurer aux économies contemporaines la flexibilité dont elles ont besoin. Dans de nombreux pays de l'OCDE, toutefois, "ce sont les étudiants, et non pas les étudiantes, qui constituent la majorité des inscrits dans les filières de technologie, de mathématique, de sciences de l'ingénieur et d'informatique ; la discrimination à l'encontre des femmes qui entrent malgré tout dans ces filières est très courante. La technologie est un important moteur de la croissance économique : il faudrait encourager les jeunes filles et les femmes à s'orienter vers les sciences exactes. [**Amy Millmann, États-Unis, Environnement économique et technologie**]

Conséquences pour l'action des pouvoirs publics et bonnes pratiques

Partenariats entre pouvoirs publics et entreprises

Les systèmes éducatifs doivent être adaptés aux besoins actuels et proposer pareillement aux hommes et aux femmes des programmes efficaces.

Ce sont généralement les pouvoirs publics qui décident de l'orientation du système éducatif formel. Le monde des entreprises intervient de plus en plus en ce qui concerne la formation des adultes et les modalités éducatives parallèles. Les gouvernements, les établissements d'enseignement et les organismes intermédiaires devraient travailler en partenariat avec les propriétaires de micro-entreprises et de PME, hommes et femmes, pour garantir une adaptation convenable des modalités éducatives formelles et informelles aux besoins du moment et leur organisation en fonction de ces différents stades de la croissance d'une entreprise. Ils devraient aussi veiller à ce que ces modalités éducatives s'adressent aux hommes et aux femmes et soient performantes dans un cas comme dans l'autre. En Australie, l'État du Queensland a créé un programme relevant de la formation à l'entreprise qui permet à l'aide de moyens financiers et humains fournis par l'État, à des élèves de l'école secondaire de rencontrer de vrais professionnels et de créer avec leur concours de vraies entreprises. [**Judith Maestracci, Australie, Accès aux marchés**]

Adapter le système éducatif au client

Le système éducatif doit être suffisamment flexible pour s'adapter à des besoins qui changent...

Les systèmes éducatifs doivent être mieux adaptés à la situation locale et au moment. Ils doivent faire appel conjointement à des méthodes et démarches d'enseignement classiques et non classiques et faire la distinction entre les besoins et les niveaux des différentes catégories d'apprenants, selon qu'il s'agit de personnes évoluant en milieu rural ou urbain, de fournisseurs de biens ou de services, d'hommes ou de femmes dirigeant une entreprise, de chefs d'entreprises qui démarrent ou sont en expansion. Les systèmes d'éducation et de formation, ainsi que la technologie associée, doivent être suffisamment flexibles pour s'adapter à l'évolution de la demande. Ils doivent adopter une approche globale pour intégrer la formation continue[3].

... et assurer aux hommes et aux femmes des résultats égaux.

Les opinions divergent sur le point de savoir si les dispositifs d'éducation et de formation visant à développer l'esprit d'entreprise ne doivent pas faire de distinction quant au sexe ou s'adresser spécifiquement à l'un ou à l'autre sexe. Ce sont en général les conditions sociopolitiques, financières et culturelles qui déterminent la démarche qui, des deux, est la plus efficace. De toute manière, les dispositifs de formation, formels et informels, doivent assurer aux unes et aux autres des résultats égaux et être pareillement ouvert aux apprenants des deux sexes.

Les hommes et les femmes apprennent différemment pour des raisons à la fois culturelles et historiques. Les femmes, enfants ou adultes, semblent apprendre différemment selon qu'elles vivent dans un milieu exclusivement féminin ou mixte et mieux réagir aux méthodes plus spécifiquement conçues pour un auditoire féminin. Nombreuses aussi sont les femmes qui préfèrent que l'enseignement soit dispensé par des femmes et sur les femmes. Certains faits paraissent indiquer aussi que les méthodes pédagogiques utilisées dans les classes mixtes pour garantir la participation des filles sont également profitables aux garçons.

Il faut proposer des femmes comme modèle aux personnes qui aspirent à créer une entreprise.

Les méthodes employées pour enseigner aux adultes à gérer une entreprise négligent en général les différences entre les sexes et proposent des modèles masculins. Il faudrait, chaque fois que l'on enseigne à être un entrepreneur, proposer des décideurs femmes comme modèle aux personnes qui aspirent à créer une entreprise. Les femmes ont souvent un autre vécu, d'autres besoins et d'autres préférences, ont fait des choix différents et connu d'autres parcours, ont quitté leur emploi et passé plusieurs années à s'occuper de leurs enfants et de leur mari avant de réintégrer les circuits économiques normaux en créant une entreprise. Elles sont souvent plus intéressées que les hommes par les techniques et les problèmes de management et moins par les questions financières. Il semblerait d'après les études faites en Australie, au Canada et aux États-Unis que les femmes aient tendance à préférer les formations peu structurées et la pédagogie expérientielle, à vouloir une femme comme professeur et comme tuteur et désirent bénéficier d'un suivi de formation. Les femmes ont apparemment surtout besoin de formation concernant la gestion financière et la technologie, ainsi que les relations commerciales et les alliances stratégiques.

Il semblerait que les femmes apprennent mieux sous la conduite d'une femme...

Comme le souligne **Candida Brush [États-Unis, Gestion]**, les femmes gèrent leur entreprise autrement ; ce sont donc les techniques utilisées par elles qu'il faut employer pour enseigner d'autres façons de gérer efficacement. Il semblerait que les femmes apprennent mieux sous la conduite d'une femme. Aux États-Unis, la National Association of Women Business Owners (NAWBO), constituée en 1975 et forte de près de 8 millions d'adhérentes, a fait pression pour que soit créé, au sein de la Small Business Administration des États-Unis, l'Office of Women's Business Ownerships (OWBO) [**Suzanne Taylor, États-Unis, Environnement économique et technologie**]. "OWBO possède 65 sites de démonstration qui proposent des formations et des services d'assistance technique aux femmes chefs d'entreprise et s'apprête à lancer sur Internet le premier site interactif de formation commerciale destiné aux femmes américaines. La technologie est ainsi un instrument de formation des femmes, particulièrement utile pour leur enseigner les questions qu'elles ont du mal à appréhender." [**Sherrye Henry, États-Unis, Environnement économique et technologie**]

... et que le tutorat soit une méthode pédagogique efficace.

Le tutorat est une méthode pédagogique efficace, bien utilisée par les femmes chefs d'entreprise qui la mettent en oeuvre. L'Australie, où a été ouvert en 1995 le Step-Ahead Business Mentoring Programme, a créé un pool de femmes et d'hommes pour servir d'exemple aux femmes dirigeant des entreprises. Cela a donné un regain de croissance et de visibilité aux entreprises dirigées par des femmes ; la quasi-totalité des 2 000 femmes qui adhèrent à ce programme ont acquis par ce biais davantage de confiance en elles-mêmes et de compétence en affaires [**Angeline Low, Australie, Mondialisation**]. L'importance de ces exemples a également été soulignée par **Zenaida Gordon [Philippines, Dialogue...]** : "Mme Lim est le chef de file des femmes entrepreneurs aux Philippines, c'est un symbole, un exemple pour chacune d'entre nous, parce qu'elle a réussi dans son projet ; elle a ouvert des coentreprises avec AIWA, Sony, Samsung, Sanyo."

Le système éducatif devrait encourager la population féminine à se servir plus systématiquement des techniques et moyens d'information.

Les structures informelles servent aussi à transmettre des informations utiles pour gérer une entreprise. Les réseaux, les associations et les relations sont indispensables pour se tenir au courant des techniques et avoir des informations sur la situation économique et les débouchés. Dans l'ensemble, les femmes tirent beaucoup moins bien d'enseignements que les hommes des meilleures pratiques internationales, nationales et locales, parce que celles-ci ne font pas la distinction entre les entreprises gérées par les femmes et celles gérées par les hommes et sont sans doute moins utiles aux femmes qu'aux hommes. Les femmes fréquentent moins les réseaux et les associations, parce qu'elles ont relativement moins de temps à consacrer à ce genre d'activités et que, n'arrivant pas à se faire une place dans ces cercles masculins, elles se découragent. Les systèmes d'information devraient faire la distinction entre entreprises féminines et masculines quand c'est utile. Quant au système éducatif, il devrait encourager la population féminine à se servir plus systématiquement des techniques et moyens d'information.

"Il faut impérativement faire connaître ces exemples de réussite aux jeunes femmes et au grand public afin que l'on sache non seulement que les femmes *peuvent* faire ces choses, mais qu'elles les *font* et qu'elles les font bien". [**Amy Millman, États-Unis, Environnement économique et technologie**]

Apprendre aux enfants à entreprendre

Plus tôt on apprend à entreprendre, mieux on apprend.

Les enfants dont les parents dirigent leur propre affaire ou ont l'esprit d'entreprise auront probablement davantage que d'autres le goût d'entreprendre. On sait par expérience que plus tôt on apprend, mieux on apprend. Le Président de l'US Export Council a créé un module multimédia intitulé "Virtual Trade Mission" pour faire comprendre aux jeunes de l'école secondaire et des premières années d'université l'importance des exportations. "Income of Her Own" est un module de formation des jeunes femmes à l'entrepreneuriat à l'aide d'études de cas réelles. Il existe aux États-Unis des camps d'été où les adolescentes apprennent à gérer une affaire [**David Bowie, États-Unis, Mondialisation**]. Dans certains établissements, on enseigne aux enfants de 8 et 9 ans à gérer un portefeuille d'actions "virtuelles" et des affaires simples.

Les visites que les enfants rendent à leur mère sur son lieu de travail sont très instructives (...) et renforcent l'image positive de la mère au travail.

Les enfants comprennent mieux comment marche une entreprise après avoir visité une usine ou une entreprise technologique (**Dina Lavoie, Mondialisation**]. Aux États-Unis, une campagne a été lancée sur le thème "Aujourd'hui, emmenez votre fille au travail" [**Cherie Piebes, Environnement économique et technologie**] qui vient renforcer l'image positive de la mère au travail et inculque aux enfants le goût d'être actif, indépendant et entreprenant. Les femmes doivent savoir qu'il existe des femmes qui ont une famille et un époux et qui sont en même temps chefs d'entreprise. Pour développer le sens des affaires chez les enfants comme chez les adultes, en milieu scolaire, universitaire ou autre, il faut enseigner l'économie, l'administration des affaires, la gestion financière, le management et les technologies de l'information.

Une formation accessible et relativement bon marché

Ce qui marche le mieux, c'est une formation facilement accessible, avec utilisation de l'informatique et mise en situation.

La formation doit être aisément accessible et bon marché et elle ne doit pas prendre trop de temps. Les femmes ont généralement une petite entreprise à faible rentabilité et ont moins de temps que les hommes à consacrer à sa gestion. Pour apprendre et se former, elles recherchent plutôt les moyens bon marché et rapides. Le plus efficace est de leur ouvrir un accès facile à des modalités d'enseignement diversifiées : dans des centres ou établissements d'enseignement, chez elles par Internet et d'autres médias technologiques, dans des entreprises moyennes à travers la mise en situation principalement, au sein de grandes entreprises par le moyen du tutorat. Le gouvernement américain a créé 53 centres dans lesquels les femmes peuvent apprendre à gérer une entreprise [**Sherrye Henry, États-Unis, Environnement économique et technologie**].

Les technologies, des outils pédagogiques essentiels

Grâce aux nouvelles technologies, on dispose d'outils pédagogiques aussi conviviaux que possible

C'est grâce à la formation que l'on apprend à se procurer les informations et les connaissances dont on a besoin. Les réseaux mondiaux d'information diffusent des programmes de formation, des renseignements, des pratiques exemplaires, des études de cas et des informations sur les marchés et les entreprises ; ce sont autant d'exemples des moyens de formation permanente dont les entrepreneurs ont besoin pour rester compétitifs.

Le boom technologique, la généralisation rapide d'Internet et des autres innovations technologiques, audiovisuelles et informatiques doivent être mis à profit pour proposer des matériels pédagogiques aussi conviviaux que possible.

Recommandations :

Un certain nombre de propositions et de recommandations concernant l'éducation et la formation ont été dégagées lors des divers ateliers et séances plénières de la conférence :

◊ *Revoir les programmes de formation des maîtres pour y inclure l'initiation à l'entrepreneuriat. Revoir les programmes d'études scolaires et d'éducation des adultes pour vérifier qu'ils comportent une initiation aux affaires, à la gestion financière, à la gestion et à la technologie. Prendre des mesures pour garantir que les matériels pédagogiques font une place équitable aux femmes.*

◊ *Les pouvoirs publics peuvent contribuer à l'établissement de partenariats éducatifs, propices à l'organisation d'ateliers d'échange d'expériences, de programmes de formation (surtout en matière de marketing, de financement et de technologie), de visites d'entreprises tenues par des femmes et d'activités d'éveil. Ces partenariats sont la clé de la réussite de tout partenariat national ou international entre établissements d'éducation et entreprises, aussi bien dans les pays Membres de l'OCDE que dans les pays non membres.*

◊ *Encourager la création de centres liés à des entreprises "volontaires" qui fourniraient les locaux, le matériel, les programmes et l'information. Veiller à ce qu'ils soient matériellement et financièrement accessibles aux femmes. Encourager les entreprises à faire don de matériels "périmés", en particulier des ordinateurs, aux centres d'orientation et de formation.*

◊ *Vérifier la rentabilité des programmes de formation et juger de l'efficacité de l'enseignement de l'entrepreneuriat imparti aux jeunes femmes et aux adolescentes à l'école.*

◊ *Engager des campagnes de promotion de l'entrepreneuriat féminin.*

◊ *Rassembler des statistiques sur l'existence et l'efficacité des programmes de formation à l'entrepreneuriat "réservés aux femmes".*

◊ *Favoriser la mise sur pied d'un réseau mondial de diffusion de l'information, des pratiques exemplaires, des études de cas et autres matériels pédagogiques. Veiller à ce que le réseau soit accessible à un maximum d'utilisateurs.*

◊ *L'OCDE devrait diffuser des informations sur les bonnes pratiques en matière de formation à l'entrepreneuriat des filles et des femmes. Elle devrait aussi faire de même en ce qui concerne les bonnes pratiques de partenariat entre établissements d'éducation et milieux de l'entreprise dans les pays de l'OCDE, les pays en transition et les pays en développement.*

C. Eliminer les Obstacles Financiers et Favoriser la Mise en Place d'Instruments Financiers Nouveaux pour Faciliter la Création et le Développement des Entreprises Appartenant à des Femmes

Principaux problèmes

Le financement des PME

La conférence a cherché à mettre le doigt sur les problèmes que les femmes rencontrent pour financer leurs entreprises et à montrer comment l'action des pouvoirs publics et des marchés pourrait leur ouvrir un meilleur accès au financement.

> "Il nous faut une stratégie à plusieurs volets : un bouquet de stratégies en direction des femmes et une stratégie pour tout le monde. Les gouvernements doivent se placer en position de chef de file, de catalyseur, s'employer à résoudre les problèmes et partager les risques. C'est la première chose. Ensuite, il faut identifier les partenaires qui pourraient participer à des investissements communs – tuteurs, investisseurs ordinaires ou organismes financiers intermédiaires, par exemple. Troisièmement, il faut inventer de nouvelles interventions et des produits financiers qui élèveront les investissements des femmes dans l'entreprise au rang de système financier complet. Enfin, il faut se servir des nouvelles technologies de télécommunication pour mettre en rapport les investisseurs et les femmes chefs d'entreprise. Les femmes qui occupent les positions de haut niveau doivent s'attacher à orienter l'action des pouvoirs publics vers le financement de tout l'éventail des entreprises, des plus petites aux plus grandes." [**Linda Tarr-Whelan, Financement**]

Les difficultés que rencontrent les femmes sont-elles inhérentes à leur condition de femme ou tiennent-elles au type de PME qu'elles créent et aux secteurs d'activité vers lesquels elles s'orientent ?

Le financement des PME est une question dont on parle beaucoup parmi les chefs d'entreprise, les banquiers, les milieux politiques et les organisations économiques nationales et internationales. Les femmes chefs d'entreprise ont au fond les mêmes caractéristiques que leurs homologues masculins et se trouvent nécessairement face à un environnement financier de plus en plus exigeant. L'accès des femmes au capital et en particulier à des fonds de démarrage est une terrible course d'obstacles, hérissée des haies des demandes de cautions, de l'argent cher et des plafonds des prêts. Les difficultés que rencontrent les femmes sont-elles inhérentes à leur condition de femme ou tiennent-elles au type de PME qu'elles créent et aux secteurs d'activité vers lesquels elles s'orientent ?

Le sentiment de différenciation ou de manque de crédibilité dont beaucoup de femmes entrepreneurs font état semble essentiellement résulter de la méfiance des banques à l'égard de certains secteurs d'activité, d'une taille d'entreprise et du manque d'expérience dans l'évaluation des besoins au démarrage et dans leur formulation financière.

Patricia Mahot [France, Financement] a souligné qu'un petit pourcentage de chefs d'entreprise obtiennent un prêt bancaire à la création, les femmes et les hommes y étant aussi nombreux. L'étude réalisée en France par l'Observatoire des entreprises en 1995 sur un panel de 30 000 créateurs a montré que 23 pour cent de femmes entrepreneurs avaient bénéficié d'un prêt bancaire, pour 22 pour cent d'hommes. A l'analyse, le sentiment de différenciation ou de manque de crédibilité dont de nombreuses femmes chefs d'entreprise font état semble plus résulter d'une méfiance à l'égard de certains secteurs d'activité, d'une taille d'entreprise et du manque d'expérience dans l'évaluation des besoins au démarrage et dans leur formulation financière. Ce que ces données enseignent en tout cas, c'est que 75 pour cent des créateurs, hommes et femmes confondus, n'ont pas bénéficié de prêts bancaires au moment de la création. Ces données demandent cependant à être mieux explorées car elles peuvent traduire soit un refus de la part des banques soit une sous-estimation des besoins ou une absence de besoins des entreprises. L'Institut autrichien de recherche sur les petites entreprises concluait, dans une étude parue en 1996, dans le même ordre d'idée, que pour 60 pour cent des hommes et des femmes chefs d'entreprise, le problème le plus important est bien celui de l'accès au financement.

Deux études conduites en Suède et en Suisse montrent qu'un homme et une femme, dans le même secteur d'activité, sont plus semblables que des femmes travaillant dans des secteurs différents. Tous les créateurs d'entreprise se heurtent à de gros problèmes de financement, mais les femmes ont plus de difficultés à les résoudre.

Comme la plupart des dirigeants ou futurs dirigeants de PME ont peu de fonds propres, ils doivent souvent recourir au crédit pour créer ou développer leur société. Le plus souvent, l'investisseur connaît mal la qualité ou les intentions de l'emprunteur, ce qui crée des problèmes de contrat. Le prêteur doit donc s'informer et établir une relation de confiance. Cette démarche a naturellement un coût fixe non recouvrable, qui est inversement proportionnel aux informations dont le banquier dispose sur l'emprunteur et à la maîtrise de la gestion que possède celui-ci. D'autre part les coûts de transaction sont sans commune mesure avec la modicité des prêts demandés. Dans l'esprit du banquier, les petites entreprises offrent peu de visibilité et de perspectives de développement, le tout dans des secteurs où l'essentiel de la valeur de l'entreprise repose sur son dirigeant. Aussi, pour le banquier, l'évaluation des risques se fait-elle souvent comme un pari.

Le financement des entreprises appartenant à des femmes

Les femmes rencontrent incontestablement des difficultés particulières sur deux plans : la réunion des fonds propres au moment du démarrage et l'accès au financement.

Malgré les caractéristiques communes qu'elles partagent avec les hommes, les femmes rencontrent des difficultés particulières pour ce qui concerne la réunion des fonds propres au moment du démarrage et l'accès au financement :

◊ La majorité des femmes créent de très petites entreprises dans le secteur des services et du commerce de détail, secteurs plutôt précaires, et opèrent en général sur des marchés de proximité. Dans de nombreux pays, le taux de défaillance des entreprises est dans le commerce de détail assez important.

◊ Les créatrices disposent de moins de fonds propres que les créateurs parce qu'elles possèdent moins d'épargne ou percevaient une rémunération inférieure dans les emplois occupés précédemment.

◊ Les femmes sollicitent dans l'ensemble moins de crédits que les hommes. Les banquiers ont tendance à assimiler ces crédits à des prêts personnels. Comme les femmes sous-estiment leurs besoins financiers ou répugnent à emprunter des sommes importantes, les banques et autres organismes ont tendance à considérer que leurs entreprises sont vulnérables. **Linda Tarr-Whelan [États-Unis, Financement]** a indiqué que selon une étude récente de la Réserve fédérale des États-Unis sur le financement des petites entreprises,

42 pour cent des dirigeantes d'entreprise se servent de leur carte de crédit personnelle lorsqu'elles ont besoin d'un crédit à court terme, 30 pour cent ont financé leur entreprise sur leur épargne personnelle, 42 pour cent des entreprises dirigées par des femmes ont moins de 25 000 dollars EU d'actifs et plus de 20 pour cent des entreprises féminines ont un chiffre d'affaires annuel inférieur à 25 000 dollars.

◊ En ce qui concerne les garanties, les femmes travailleurs indépendants ou chefs d'entreprise sous statut juridique spécifique répugnent souvent à engager le patrimoine familial dans leur activité professionnelle. Dans certains cas, l'accord du mari ou de la famille est exigé pour l'obtention d'un prêt. Aux mesures bancaires discriminatoires qui existent dans certains pays, s'ajoutent l'inégalité des droits devant l'héritage, l'inégalité d'accès des femmes au patrimoine en cas de divorce, l'absence de statut des femmes collaboratrices de leur mari qui leur interdit de faire valoir leurs droits à l'entreprise (d'où la perte éventuelle des droits à la retraite ou aux prestations de sécurité sociale), etc.

◊ Les femmes ont souvent des problèmes aigus en matière d'information et de relations publiques. **Fotini Legaki [Grèce, Financement]** a souligné combien il était difficile pour une femme de nouer des relations personnelles solides avec son banquier et fait observer que les dirigeantes de PME étaient souvent mal informées de l'existence de nouveaux instruments financiers – crédits-bail, affacturage, capital-risque, etc. **Dina Lavoie [Canada, Financement]** a indiqué que rares sont les femmes qui réussissent à bénéficier des subventions officielles, le plus souvent parce qu'elles sont informées trop tard de leur existence – "les courriers ne leur sont pas adressés". On ne demande presque jamais à une femme de siéger dans les commissions des banques de développement qui accordent des prêts aux entreprises. Les formalités qu'exige toute transaction bancaire sont une énorme perte de temps pour une petite entreprise. **Naoko Banno [Japon, Gestion]** a signalé que les difficultés de la procédure avaient empêché son entreprise de bénéficier d'une subvention pour le développement d'un logiciel.

◊ Comme les femmes sont arrivées plus tardivement que les hommes sur le marché du travail dans certains métiers, elles ont souvent moins d'expérience professionnelle, en particulier en matière de direction ou d'encadrement et d'accès au financement.

> "Il faut en tout premier lieu faire tout ce qui est en notre pouvoir pour que les femmes disposent d'un revenu plus élevé. L'accès au crédit bancaire est important, mais la capacité d'accumuler des fonds propres l'est tout autant." [**Leila Webster, Banque Mondiale**]

Conséquences pour l'action des pouvoirs publics et bonnes pratiques

Les dispositifs de financement s'adressant aux micro-entreprises faciliteront vraisemblablement la tâche d'un grand nombre de femmes chefs d'entreprise.

Il serait très intéressant de savoir quelle est l'ampleur de la réaction du marché à cet ensemble de phénomènes, s'il y en a une et, s'il n'y en a pas, quels dispositifs publics sont mis en place pour faciliter l'accès des entreprises dirigées par des femmes au financement. On sait que des organismes financiers ont mis en place des mécanismes de crédit spécifiquement destinés aux femmes avec dans certains cas des résultats remarquables. Ainsi, **François Beaudoin [Canada, Financement]** a indiqué que "50 pour cent des créations, souvent de micro-entreprises, sont en fait dues à des femmes. La Banque fédérale du Canada prévoit que d'ici à dix ans, compte tenu du rythme de créations d'entreprises par des femmes, il y aura au Canada égalité entre le nombre d'entreprises dirigées par des hommes et par des femmes. Il y a là un gisement de clientèle important qui devrait intéresser tous les établissements financiers".

On peut donc escompter que les différents dispositifs aptes à répondre convenablement aux demandes de financement de micro-entreprises ou de très petites entreprises faciliteront la tâche d'un grand nombre de femmes entrepreneurs.

D'autres systèmes agissent dans le sens d'une plus grande équité en faisant la soudure entre les financements familiaux et amicaux et les apports de capital-risque.

Henry Savajol [France, Financement] a fait état de l'existence en France de dispositifs qui, au nom de l'équité, visent à faire la soudure entre les financements amicaux et familiaux et les apports de capital-risque. La Banque de développement des PME a récemment proposé des mesures destinées à encourager les investissements individuels directs dans des PME non cotées, à travers des fonds de placement spécialisés, partiellement garantis par l'État, dans une de ses filiales spécialisées, la SOFARIS (Société française de garantie de financement des PME). Des possibilités analogues existent déjà pour les sociétés de capital-risque, ainsi que pour mettre en place des procédures de placements en obligations assimilées à des actions en ce qui concerne les petites entreprises en phase de développement. Cette formule est inspirée des prêts participatifs créés en France dans les années 80 et des prêts à redevances de la Banque fédérale de développement du Canada. Il s'agit de prêts à six ans non garantis à intérêt faible et redevances calculées sur le chiffre d'affaires de l'emprunteur.

Les fonds de garantie fournissent les assurances nécessaires pour faciliter les prêts bancaires...

Henry Savajol a également souligné que pour accorder des prêts aux micro-entreprises à faible potentiel de croissance, les banques devaient avoir des assurances que les chefs de ces entreprises ne peuvent pas leur fournir eux-mêmes. Les fonds de garantie sont parfaitement adaptés à de telles situations. Il existe en France un certain nombre de dispositifs de garantie : fonds privés (organismes de garantie mutuelle) ; fonds publics administrés par la SOFARIS et le Fonds de garantie aux initiatives féminines en France (FGIF, créé en 1989) auquel peuvent s'adresser les femmes qui souhaitent créer, acheter ou développer une entreprise (le montant maximum des prêts fixé est à 150 000 francs français et la garantie s'étend à 60 pour cent du prêt).

Nancy Barry [États-Unis, Dialogue ...] a fait observer que l'organisme Women's World Banking avait su développer le concept de capitalisation sous forme de dotation en capital pour les établissements de micro-financement des pays en développement. Il s'agit de se constituer son propre portefeuille et de prêter, puis de se servir, de ce potentiel pour mobiliser des fonds auprès des banques locales. Un concept analogue pourrait être mis au point pour les organismes de développement des entreprises de manière à améliorer l'accès des femmes au financement et d'apporter ainsi des solutions aux problèmes généraux de développement économique.

... tandis que d'autres programmes permettent aux femmes d'obtenir une lettre de pré-qualification avant de présenter à la banque leur dossier d'obtention d'un prêt...

Sherrye Henry [États-Unis, Environnement économique et technologie] a expliqué que la Small Business Administration (SBA) avait mis en place un programme de micro-prêts (500, 5 000 et 25 000 dollars EU) s'adressant aux femmes. Un autre dispositif, dit documentation des prêts, permet aussi à la SBA de garantir les emprunts bancaires des femmes à hauteur de 100 000 dollars. Un autre dispositif encore de la SBA permet aux femmes, particulièrement celles qui ont des entreprises de services, d'obtenir une lettre de pré-qualification avant de présenter à une banque leur dossier d'obtention d'un prêt.

...et d'autres encore associent prêts, formation et activités de conseil.

D'autres dispositifs associent prêts, formation et activités de conseil. La Banque fédérale de développement du Canada en a plusieurs. Ainsi que l'a expliqué **François Beaudouin [Canada, Financement]**, la Banque a lancé, il y a trois ans, le programme Step-up, programme de tutorat de créatrices d'entreprise par des entrepreneurs plus expérimentés. Depuis son ouverture, 592 tutorats ont été mis en place qui ont donné de bons résultats ; la corrélation est forte entre l'amélioration des aptitudes à la gestion et la performance financière. M. Beaudoin a également précisé que la BDC accordait 30 pour cent de ses prêts à des femmes ou à des entreprises comptant au moins une femme parmi ses actionnaires majoritaires. La banque a constaté que le pertes étaient moindres sur les prêts accordés aux femmes. Elle a aussi mis en place un dispositif s'adressant aux micro-entreprises qui associe formation, conseil et financement et dont elle se dit satisfaite dans cette phase de démarrage. Sur une période de 12 mois (1996-97), 384 prêts ont été approuvés, pour un montant total de 12 millions de dollars canadiens. Pour l'instant, 30 pour cent seulement des bénéficiaires sont des femmes, mais l'on s'attend à ce que le pourcentage augmente nettement. Sur le total des dirigeantes qui ont suivi le programme, 32 pour cent avaient une entreprise à forte intensité de savoir ou tournée vers l'exportation, 22 pour cent un commerce et 18 pour cent une activité relevant du secteur manufacturier.

Certains établissements bancaires prêtent aux jeunes entreprises qui ne remplissent pas les conditions requises pour obtenir un prêt classique...

Il y a quatre ans, une grande banque hollandaise, la Banque ABN AMRO a ouvert un département s'adressant aux créateurs d'entreprise, dont 20 à 25 pour cent de la clientèle sont des femmes. Selon cette banque les prêts accordés aux femmes sont inférieurs de près des deux tiers à ceux accordés aux hommes parce que les femmes ont tendance à créer des entreprises qui fonctionnent à temps partiel et ont nécessairement un chiffre d'affaires et des besoins de financement moindres. **Robert Hughes [Irlande, Financement]** a donné l'exemple de la Banque d'Irlande qui dispose d'un fonds d'environ 60 millions d'écus pour des prêts affectés au financement de créations d'entreprises qui ne satisfont pas aux critères bancaires classiques. Cette banque est aussi partie prenante d'un programme avec First Step, pour associer financement et conseil à des petites entreprises en phase de démarrage.

...tandis que le " Loan Support Scheme for Women" en Suède est un concept nouveau, mis en place par l'État, pour aider les femmes entrepreneurs à créer de nouvelles

Il existe aussi des dispositifs publics de soutien de caractère national et international. **Anne Löfquist [Suède, Financement]** a signalé que le "Loan Support Scheme for Women" en Suède était un concept nouveau mis en place tout exprès par l'État pour aider les femmes entrepreneurs aussi bien pour la création d'entreprise que pour le développement *entreprises ou à développer des entreprises existantes...*

d'entreprises déjà existantes. Outre un prêt qui pourrait s'élever à 15 pour cent du capital requis (d'un montant compris entre 3 000 et 17 000 écus), le programme organise des actions de tutorat, des stages, des séminaires, des formations pour membres de conseils d'administration, etc. Depuis son démarrage en 1994, 5 600 emplois ont été créés et plus de 3 000 entreprises ont emprunté plus de 26 millions d'écus. Quarante pour cent des prêts vont à des entreprises de services, 50 pour cent à des commerces (de détail) et 8 pour cent seulement à des entreprises du secteur manufacturier. La durée des prêts est de six ou neuf ans ; aucun intérêt ni frais d'amortissement n'est payable avant deux ou trois ans. Selon Anne Löfquist, ce dispositif a largement incité les femmes à créer ou développer leur propre entreprise.

... et la Commission européenne s'est occupée de la création d'entreprise par les femmes, y compris des aspects financiers, depuis plus de dix ans.

Hella Gerth [Commission européenne, Financement] a expliqué que la Commission s'intéressait aux créatrices d'entreprises depuis 1986 dans le cadre de ses programmes sur la création d'emplois. Le programme d'initiatives locales pour l'emploi, puis le programme NOW (New Opportunities for Women) obtiennent de bons résultats et donnent des exemples positifs d'entreprises créées par des femmes. Après évaluation de ces exemples, il a été décidé de proroger le programme NOW jusqu'en 1999. La dimension féminine est prise en compte dans les principaux programmes de soutien de l'Union européenne. La Commission dispose d'un petit budget pour organiser des actions pilotes, des séminaires et des conférences ouvertes à tous les chefs d'entreprises petites et micro, sans distinction de sexe.

La Banque Grameen fut créée par le Dr. Yunus en 1983. Il révolutionna ainsi le prêt aux plus pauvres en développant une "technologie" fondée sur de fortes incitations au remboursement (groupement de prêts avec obligations mutuelles de remboursement), rapidité, suivi attentif avec supervision locale et prêts renouvelés. Grâce à ce travail de pionnier : plus de 20 millions de personnes, presque exclusivement des femmes (à 94 pour cent) ont été capables de rompre avec la pauvreté extrême par la création de micro-entreprises. Au Bangladesh, 10 pour cent de la population (10 millions de personnes) bénéficient de la Banque Grameen. Le modèle de la Banque Grameen, avec son taux de remboursement de 90 pour cent, a été reproduit avec succès et intégré aux divers dispositifs de prêts publics et privés de plus de 56 pays de l'OCDE et pays en développement, comme le Canada, la Chine, la Norvège, l'Afrique du Sud et les Etats-Unis (notamment en Arkansas et à Chicago).

Recommandations :

Il importe de mieux connaître les entreprises dirigées par des femmes, en particulier les très petites entreprises (modes de financement des créations, épargne personnelle, prêts familiaux, prêts bancaires, montant moyen des prêts, pourcentage des demandes de prêt acceptées, pourcentage des demandes rejetées, etc.). Il importe aussi de juger de la pertinence et de l'efficacité des "bonnes pratiques". Sont-elles des "pratiques exemplaires" ? Doivent-elles être diffusées de manière à encourager le recours à des produits et instruments financiers éprouvés ?

*Le débat sur ces questions a été très animé, en particulier sur le point de savoir s'il fallait promouvoir des actions, dispositifs ou mesures de soutien spécifiques : "les programmes spécifiques s'adressant aux femmes sont-ils discriminatoires ou constituent-ils une action positive ? Il nous faut des actions positives pour remédier aux disparités, mais il existe maintenant dans chaque pays une loi sur l'égalité, une loi antidiscrimination. Le Loan Support Scheme for Women suédois serait inacceptable dans de nombreux pays" **[Robert Hughes, Irlande, Financement]**. "Personnellement, je voudrais plaider pour une chose : pas de distinction de sexe dans ces recommandations, car elles sont tout aussi importantes pour les hommes qui dirigent une entreprise que pour les femmes qui dirigent une entreprise" **[Ian Hamilton Fazey, Royaume-Uni, Financement]**. Le sentiment généra, toutefois, a été qu'apporter des solutions aux entreprises dirigées par des femmes serait profitable à l'ensemble des petites entreprises.*

Les participants à la conférence ont également reconnu qu'il importait de mettre en regard les objectifs des investisseurs, en termes de critères d'investissement et de besoins en capital, et les besoins des petits entrepreneurs soucieux d'assurer le développement de leur affaire à différents stades de son existence. Cette démarche pourrait permettre de repérer les manques et de trouver de nouveaux partenaires pour les combler.

Les participants sont convenus que si l'on voulait faciliter l'accès au crédit des micro-entreprises en général et des entreprises dirigées par des femmes en particulier, ainsi que la pérennité de ce crédit, il fallait que les prêteurs ressentent moins le "risque" qu'il comportait. Les démarches ci-après sont proposées aux responsables politiques et décideurs pour mise en oeuvre éventuelle après adaptation aux conditions nationales ou locales :

◊ *Faire en sorte que les femmes chefs d'entreprise disposent d'un volume plus important de fonds propres en mettant en place des programmes d'épargne, d'accès à la propriété du logement, etc. ;*

◊ *promouvoir la mise en place de dispositifs appropriés de garantie des prêts;*

◊ *mettre un pool de crédit à la disposition des femmes dirigeant une entreprise de taille quelconque sur un "second marché" des prêts aux petites entreprises ; créer des programmes de financement des micro-entreprises ;*

◊ *ouvrir plus largement l'accès au crédit en favorisant la création d'organisations intermédiaires, structures intermédiaires entre les grands investisseurs et les très petits emprunteurs (banque Grameen), qui peuvent consentir des prêts d'un faible montant à un coût raisonnable ;*

◊ *faire progresser le taux de réussite des entreprises féminines en encourageant les services de conseils aux femmes qui créent ou développent une entreprise et en ouvrant à celles-ci l'accès aux réseaux de soutien concernant la formation, l'information, la recherche des marchés, ainsi que la protection sociale (assurance maladie) et la garde des enfants ;*

◊ *développer des partenariats éventuels entre les pouvoirs publics et les ONG et les établissements de prêt privés afin de rendre ce marché attractif ;*

◊ *encourager les organisations internationales, telles que l'OCDE, à approfondir les connaissances sur le financement des entreprises appartenant à des femmes et à diffuser les pratiques exemplaires dans ce domaine dans la zone de l'OCDE, ainsi que dans les pays en transition et en développement.*

D. Promouvoir la Compétitivité des PME dirigées par des Femmes

Principaux problèmes

Un monde dominé par le savoir, les réseaux et la communication interactive, les technologies nouvelles, l'innovation et les préoccupations sociales (...) et un nombre croissant de femmes chefs d'entreprise.

Le monde change et le milieu dans lequel évolue l'entreprise change lui aussi. Les futurologues prédisent que le XXIème siècle sera dominé par le savoir, les réseaux et la communication interactive, les technologies nouvelles, l'innovation et les préoccupations sociales (comme de maintenir le niveau de vie et de parvenir à se faire reconnaître comme individu). Dans cet univers de mutations économiques et d'incertitudes, qui obligeront inévitablement les entreprises à être de plus en plus flexibles, à s'organiser en réseau et à être plus attentives aux besoins humains, les PME présentent de nombreux avantages. Le XXIème siècle verra aussi sans aucun doute les femmes changer de rôle et le nombre de celles qui dirigent une entreprise dans le monde s'accroître, autre source probable de changement et d'innovation dans les milieux de l'entreprise et la société. **Thierry Gaudin [France, Environnement économique et technologie]** a fait remarquer que les changements technologiques ont conduit à de nouvelles formes d'organisation du temps et de nouveaux modes de communication qui ressemblent peut-être davantage à la manière féminine de "faire les choses".

Pour préserver leur compétitivité, les PME doivent accroître leurs investissements matériels et immatériels et leur capacité stratégique (...), d'où la nécessité d'une gestion appropriée pour localiser et développer les ressources nécessaires.

La compétitivité des PME repose sur toute une série de facteurs : *a)* les investissements matériels dans un équipement du niveau technologique approprié ; *b)* des investissements immatériels dans toute une série de domaines – aptitude à se servir efficacement des outils technologiques et des logiciels ; capacité d'exploiter les informations sur les marchés et la concurrence ; niveau de formation des personnels ; qualité de l'organisation de l'entreprise ; capacité de R-D et contact avec les instituts de recherche ; *c)* des capacités stratégiques, à savoir innovation et flexibilité. Les stratégies concurrentielles des PME sont très disparates ; elles dépendent du type d'entreprise, du secteur d'activité, de sa situation géographique, de son marché, du cadre réglementaire et de la situation économique. D'où la nécessité d'une bonne gestion pour pouvoir localiser et développer les ressources nécessaires.

Quelle est l'incidence de ces facteurs sur les PME féminines ?

Il faut analyser sur cette toile de fond la question de la capacité des femmes à ouvrir et développer une entreprise et à la rendre concurrentielle. Il est ressorti des informations données au cours des différents ateliers et des débats que l'on connaissait encore assez mal l'incidence précise de ces facteurs sur les entreprises dirigées par des femmes. Les participants à plusieurs ateliers ont cependant longuement débattu de l'importance de la gestion et de l'utilisation des outils technologiques et des réseaux ; cela a été en particulier le cas dans l'atelier 1 sur l'environnement économique et la technologie, et l'atelier 3 sur le Management.

Les méthodes de gestion des femmes et les structures des entreprises qu'elles dirigent ont apparemment un caractère particulier

Il semble que les femmes structurent leur entreprise autrement, avec une organisation

L'organisation et la direction du personnel sont des facteurs cruciaux de la compétitivité. Selon **Candida Brush [États-Unis, Gestion]**, un homme et une femme n'utilisent pas les mêmes moyens pour atteindre un *hiérarchique plus resserrée et en recherchant beaucoup plus le consensus...*

même résultat. Les femmes apparemment structurent leur entreprise autrement, avec une organisation hiérarchique plus resserrée et en recherchant davantage le consensus. On voit aussi d'après l'exemple des États-Unis que les femmes sont plus disposées à engager un personnel diversifié et qu'elles offrent à leurs salariés une palette différente de bénéfices (comme par exemple le partage du travail, le remboursement des frais de formation, etc.) Les femmes entrepreneurs contribuent davantage au bien-être social de la collectivité, privilégient, semble-t-il, l'efficacité organisationnelle et sociale et sont plus soucieuses de communication interne que leurs homologues masculins. Ces différences, auxquelles s'ajoutent la capacité d'écoute des femmes, leur souci de maintenir autour d'elles un climat d'harmonie, devraient les aider à s'adapter aux exigences nouvelles qui seront celles des chefs d'entreprise au XXIème siècle. Comme l'a déclaré **Swanee Hunt** [**États-Unis, Dialogue ...**], "les femmes sont le ciment de la société. Elles savent à partir de leur propre expérience créer des liens avec autrui. L'esprit critique que possèdent la plupart des femmes est une qualité primordiale pour qui veut entreprendre, une qualité dont il faut encourager les femmes à se servir."

... mais elles semblent aussi avoir moins de temps à consacrer à leur entreprise et répugnent davantage à prendre conseil.

D'autres facteurs ont un lien indirect avec la gestion d'une entreprise, par exemple le fait que les femmes aient moins de temps à y consacrer. En effet, dans les pays de l'OCDE comme ailleurs, ce sont toujours les femmes qui assument l'essentiel de la responsabilité des charges familiales, des soins du ménage et des enfants – autant de choses qui réduisent le temps qu'elles ont à consacrer à une affaire. Les femmes ont aussi tendance à moins vouloir consacrer du temps à prendre conseil lorsqu'elles créent ou développent une entreprise, souvent parce que les services en question ne sont pas orientés spécifiquement vers les femmes, se situent dans des contextes très "masculins" et ne répondent pas aux difficultés spécifiques que rencontrent les femmes. Les femmes entrepreneurs ont par conséquent tendance à avoir une attitude extrêmement pragmatique à l'égard des outils technologiques économiseurs de temps.

Les outils technologiques sont la clé de la compétitivité

Les technologies nouvelles compensent la petitesse et la distance et permettent aux entreprises de croître et de se mondialiser...

Les technologies et les systèmes d'information changent profondément la manière dont les petites entreprises, et spécialement celles dirigées par des femmes, travaillent. La technologie intervient de plus en plus à tous les niveaux de la compétitivité : produits ; techniques de production ; méthodes de gestion ; organisation de l'entreprise ; formation du personnel ; information sur les marchés. **Ingrid Andersson** [**Suède, Mondialisation**] a expliqué que les technologies nouvelles compensent la petitesse et la distance et permettent aux entreprises de croître et de se globaliser.

... et changent la façon dont les femmes travaillent.

La technologie change surtout la manière dont elles exécutent et gèrent leur travail. **Andrina Lever** [**Canada, Mondialisation**] a fait observer que la conférence électronique, Internet, le commerce électronique (qui peut présenter une entreprise de façon extrêmement neutre), les réseaux électroniques (pages d'accueil) et le travail à domicile au moyen d'Internet sont quelques unes des innovations technologiques majeures qui permettent aux femmes de créer et de développer leurs entreprises et d'en garantir la compétitivité. La sous-traitance et le télétravail pourraient bientôt rendre totalement anachronique le fait de devoir chaque matin se rendre à son travail.

Les entreprises appartenant à des femmes , tout autant que les entreprises en général, utiliseront de plus en plus le réseau internet comme moyen indispensable pour accéder à l'information et aux opportunités d'affaires.

En l'an 2000, il y aura environ 200 millions d'internautes. Internet devient un moyen privilégié d'accès à l'information sur les normes de qualité, les sujétions légales et réglementaires, la réglementation fiscale et les débouchés, ainsi qu'un instrument pour s'informer, apprendre à utiliser des pratiques exemplaires et pour savoir qui contacter. **Kaori Sasaki [Japon, Environnement économique et technologie]** a expliqué que les femmes qui dirigent une PME doivent souvent surmonter plusieurs obstacles simplement pour se positionner sur la ligne de départ au même moment que leurs homologues masculins. Internet aide les femmes à se placer plus vite, à rester compétitives ; il aide de même les personnes qui soufrent d'un handicap physique ou qui vivent dans une région écartée. **Judith Maestracci [Australie, Accès aux marchés]** a fait observer : "il y a littéralement des milliers d'associations de femmes entrepreneurs. Il n'y a aujourd'hui aucun moyen de les repérer toutes. Je crois qu'Internet peut rassembler toutes ces informations, ce qui permettra aux femmes de tous les pays d'y accéder, de connaître les exemples, de savoir ce qui se passe, où les meilleures pratiques sont mises en oeuvre, etc. Le réseau, en plus du reste, est un impératif".

Conséquences pour l'action des pouvoirs publics et bonnes pratiques

La baisse des prix des télécommunications est un avantage pour les entreprises, mais à technologies nouvelles, compétences nouvelles.

La déréglementation mondiale des télécommunications profitera aux petits entrepreneurs en faisant baisser le coût relativement élevé dans certaines régions des communications par Internet et autres télé-réseaux. Cependant, utiliser le Net et pratiquer le commerce électronique nécessitent de la dextérité, une adaptation et des investissements matériels – là encore la formation et l'éducation sont absolument indispensables. Mais, comme **Jocelyne Attal [États-Unis, Environnement économique et technologie]** l'a expliqué, les femmes n'ont pas toutes à leur disposition un ordinateur personnel leur permettant de commercer sur le Net, surtout dans les pays en développement et les économies en transition. Aussi faut-il mettre en place des réseaux d'ordinateurs et des postes de travail Internet. De plus, le Net n'est pas nécessairement aussi propice aux affaires qu'il pourrait l'être. Les concepteurs et les prestataires de services devraient s'efforcer de rendre le réseau plus convivial et plus propice aux affaires.

Internet donne des informations sur les pratiques exemplaires du monde entier...

Sherrye Henry [États-Unis, Environnement économique et technologie], Judith Maestracci [Australie, Gestion et Arline Wouterz [Royaume-Uni, Accès aux marchés] ont fait observer qu'une page d'accueil internationale diffusant des informations sur les pratiques exemplaires et des études sur des entreprises dirigées par des femmes ou appartenant à des femmes serait extrêmement utile. L'Association *Femmes chefs d'entreprises mondiales* (FCEM) possède ainsi sa page d'accueil et ses organisations adhérentes dans 33 pays ont toutes une page d'accueil reliée à celle de la FCEM. **Etta Carignani [Italie, Dialogue ...]** a fait valoir que cela leur permettait d'être interactives ; ces femmes possèdent maintenant une liste de marketing d'environ 45 000 membres qui sont reliées entre elles et font des affaires. La FCEM est en relation avec la NAWBO et, comme l'a signalé **Suzanne Taylor [États-Unis, Environnement économique et technologie]**, des membres de l'association ont effectué des missions commerciales à Francfort, à Madrid, au Mexique, au Portugal et à Tokyo.

... et les réseaux électroniques des informations sur les investissements et les commandes possibles.

Acenet est un réseau électronique américain s'adressant aux investisseurs disposés à mettre de l'argent dans des entreprises appartenant à des femmes ; Pronet est, de même, un réseau américain qui permet aux femmes chefs d'entreprise qui y entrent des informations sur leur société de postuler pour l'attribution de marchés publics. **Sherrye Henry [États-Unis, Environnement économique et technologie]** estime que ces deux nouveautés technologiques aideront grandement les femmes américaines et pourraient servir de modèles à d'autres pays.

Toutefois, les réseaux féminins devraient être renforcés ...

Ces exemples montrent bien le lien entre compétitivité, emploi des technologies électroniques et réseaux, tant réels qu'électroniques. Cependant, comme l'a souligné **Lorraine Martin**, les femmes sous-estiment souvent l'intérêt des réseaux. **Sonia de Avelar [Brésil, Dialogue ...]** a insisté sur ce point en ces termes : "il faut renforcer les associations et les réseaux de femmes, de manière surtout à lier les associations locales et nationales entre elles à travers les réseaux internationaux, et faire un effort nord-sud de création de coalitions pour élargir l'accès collectif des femmes aux ressources économiques et leur donner davantage de moyens d'action".

... et les femmes devraient participer davantage à la conception des outils technologiques.

Si les femmes manient les outils technologiques dans leurs entreprises, peu nombreuses sont celles qui inventent ou conçoivent des équipements, des machines ou des technologies ou même donnent leur avis sur leur conception, a signalé **Helen Appleton [Royaume-Uni, Environnement économique et technologie]**. Les femmes devraient prendre contact avec les fabricants, les ingénieurs et les concepteurs de manière à faire prendre en compte leurs besoins, leurs points de vue et leurs aspirations spécifiques dès le processus de conception, aujourd'hui trop souvent coupé des utilisateurs et surtout des plus petits. **Amy Millman [États-Unis, Environnement économique et technologie]** a instamment demandé aux femmes de s'investir dans les problèmes technologiques si elles veulent rester compétitives. Se servir des outils technologiques n'est pas assez, les femmes doivent créer la technologie et diriger des équipes de R-D.

Recommandations :

Les participants à la conférence ont formulé un certain nombre de propositions et de recommandations de politique générale :

◊ *Continuer les recherches sur les liens entre avancées technologiques et ouverture de débouchés pour les PME en général, et notamment celles dirigées par des femmes.*

◊ *Faciliter l'accès aux technologies et les mettre largement à la disposition des entrepreneurs et spécialement des femmes chefs d'entreprise, des femmes disposant de peu de revenu et des mères célibataires. Ouvrir aux PME l'accès aux technologies avancées et les former en conséquence. Veiller à inscrire les savoir-faire technologiques et l'enseignement des techniques d'accès à l'information dans les politiques de l'éducation.*

◊ *Tenir compte des besoins, des souhaits et des exigences des femmes entrepreneurs lors de la formulation des programmes relatifs aux technologies de l'information et aux réseaux.*

◊ *Promouvoir la constitution d'alliances et le tutorat entre associations, entreprises et pouvoirs publics. Multiplier et développer les associations et les réseaux qui aident les femmes chefs d'entreprise à se servir des technologies pour améliorer leur compétitivité.*

◊ *Entreprendre des recherches comparatives sur les questions de gestion en analysant les pratiques partagées et les enseignements à tirer et en rassemblant des statistiques.*

◊ *Définir, au sein d'organisations internationales telles que l'OCDE, les pratiques exemplaires à suivre en matière d'utilisation de la technologie aux fins de la compétitivité et diffuser des pratiques exemplaires concernant les conseils, les appuis et la formation à assurer aux femmes qui souhaiteraient pénétrer sur le marché émergent des entreprises de haute technologie.*

◊ *Etablir, au sein d'organisations internationales telles que l'OCDE, des recommandations modèles à l'adresse des gouvernements et de leurs organismes d'aide quant aux caractéristiques de la gestion féminine, à leur pertinence par rapport aux exigences économiques et sociales du XXIème siècle et à l'adaptation possible des programmes de formation aux spécificités des femmes chefs d'entreprise.*

E. Faciliter l'Accès aux Marchés Internationaux et Intérieurs

Principaux problèmes

La mondialisation des activités économiques (...) permet aux entreprises de pénétrer sur de nouveaux marchés, de tirer parti de leurs avantages technologiques et organisationnels et de réduire les coûts et les risques.

Bien que la flexibilité des PME leur offre de nombreux avantages en leur permettant de s'adapter à l'évolution des marchés, l'accès aux marchés reste problématique dans un environnement compétitif qui n'oeuvre pas toujours en leur faveur. Aujourd'hui, la mondialisation des activités économiques, qui implique une évolution des activités transfrontières des entreprises – investissement international, échanges et alliances stratégiques, production, approvisionnement et marketing – permet aux entreprises de pénétrer sur de nouveaux marchés, de tirer parti de leurs avantages technologiques et organisationnels et de réduire les coûts et les risques. Le progrès technologique, la libéralisation des marchés et la mobilité accrue des facteurs de production sous-tendent cette expansion et sont en partie induits par elle.

Les PME appartenant à des femmes commencent tout juste à se développer à l'exportation...

Habituellement, les PME ne sont pas associées à l'activité internationale mais, dans les faits, elles jouent déjà un rôle important au plan mondial, produisant directement 26 pour cent environ des exportations de l'OCDE et quelque 35 pour cent des exportations de l'Asie. En ce qui concerne les PME dirigées par des femmes, les données laissent à penser que leur degré de mondialisation est très limité. Les femmes ne font que commencer à développer leurs exportations. Toutefois, comme l'a souligné **David Bowie [États-Unis, Mondialisation],** on constate aux États-Unis qu'en général les entreprises dirigées par des femmes et présentes sur le marché des exportations sont mieux implantées et ont un volume de ventes supérieur, un plus grand nombre de salariés et de meilleures perspectives de croissance que les entreprises qui ne font pas de commerce international.

... mais les nouvelles formes d'activités économiques mondiales pourraient permettre à plus de PME – et notamment celles dirigées par des femmes – d'être plus compétitives au niveau international.

Globalement, l'évolution survenue dans les communications, la technologie et l'organisation de la production a encouragé le processus de mondialisation. Un plus grand nombre de PME, y compris de PME dirigées par des femmes, peuvent soutenir avec succès la concurrence au plan international. **Andrina Lever [Canada, Mondialisation]** a fait observer que les PME étaient mieux à même de faire du commerce au plan international grâce à la technologie, au partenariat, au franchisage, aux accords de licence, au marketing en réseaux, aux distributeurs étrangers, aux alliances sociales, aux coentreprises et à l'approvisionnement d'autres entreprises ou de pays à titre de sous-traitants directs ou indirects. Ceux qui interviennent dans les créneaux spécialisés et qui font preuve de vivacité, de rapidité et d'imagination, qui sont à l'écoute de la clientèle et qui possèdent des compétences interactives en matière de marketing leur permettant d'envisager et de créer de nouveaux marchés seront récompensés. Il semble dans ce cas que les femmes entrepreneurs soient bien placées.

> "Les multinationales emploient rarement des femmes à la tête de leurs divisions internationales. En outre, en France, les PME ne représentent que 20 pour cent des échanges internationaux. Il est donc vital que les femmes puissent acquérir plus facilement une formation et une expérience approfondies dans le domaine des affaire internationales."[**Christine Chauvet, Directeur général, Centre français du commerce extérieur, France]**

Les migrations encouragent l'internationalisation des entreprises...

Andrina Lever a indiqué que le processus de mondialisation était également encouragé par le phénomène mondial des vagues migratoires qui se sont amplifiées ces dernières années. L'accroissement des migrations crée des communautés ethniques qui maintiennent des liens culturels et des contacts personnels avec leurs pays d'origine, et des groupes ethniques dans les pays d'accueil qui mettent à profit les débouchés commerciaux.

... mais les obstacles aux échanges et à l'investissement réduisent l'accès aux marchés.

L'accès aux marchés reste un problème pour de nombreuses PME internationalisées. Souvent, ces entreprises se heurtent à des obstacles aux échanges et à l'investissement, en particulier dans les pays en développement. Outre les obstacles tarifaires et les réglementations douanières, les obstacles non tarifaires, de même que de nombreux autres facteurs – procédures locales d'approbation, normes, coût de la création d'une société, difficultés liées à l'application de mécanismes d'organisation flexibles (accords de franchise), obstacles culturels aux réseaux de distribution ou aux ressources – risquent de limiter l'accès au marché de façon intentionnelle ou non. A ce sujet, **Ingrid Andersson [Suède, Mondialisation]** a signalé que les traits spécifiques des femmes entrepreneurs au regard des relations humaines et de l'aptitude à instaurer la confiance pouvaient conditionner leur entrée sur les marchés étrangers. Elle a fait observer qu'en fonction des circonstances, l'interaction entre les caractéristiques des secteurs industriels et la spécificité des pays et des cultures pouvait être la source aussi bien d'avantages et de possibilités que d'obstacles et de difficultés dans les échanges internationaux.

Des données laissent à penser que les femmes se heurtent à des obstacles spécifiques, dus en partie à la persistance de pratiques discriminatoires et à leurs responsabilités familiales.

Les PME sont confrontées à de nombreux défis qui sont encore plus difficiles à relever par les PME dirigées par des femmes lorsqu'elles veulent exporter. Leur principal obstacle est souvent lié à leur taille – ces entreprises n'atteignant pas généralement une taille à partir de laquelle il devient rentable de s'internationaliser – à leur activité et à leur accès à l'information. Des données laissent toutefois penser que les femmes se heurtent à des obstacles spécifiques dus en partie à la persistance de pratiques discriminatoires et à leurs responsabilités familiales. **Angeline Low [Australie, Mondialisation]** a fait observer que si les femmes considèrent qu'elles ont des compétences particulières qui leur donnent un atout pour traiter avec des représentants des entreprises et des pouvoirs publics dans d'autres pays, elles doivent trouver un équilibre entre leur activité professionnelle et leurs responsabilités familiales, et tenir compte des différences culturelles de comportement qui peuvent leur poser des problèmes particuliers dans de nombreux pays. En outre, les participants à la conférence ont souligné que les femmes entrepreneurs connaissaient des difficultés spécifiques lorsqu'elles cherchaient à financer leurs projets, dans la mesure où de nombreuses femmes entrepreneurs choisissaient d'exercer leur activité, du moins initialement, dans le secteur des services ou du commerce de détail que les institutions financières sont peu disposées à financer. Comme l'a suggéré **Uma Reddy, [Inde, Dialogue...]**, un mécanisme de garantie du crédit établi par l'intermédiaire des banques internationales agissant en qualité de garants auprès des banques locales nationales pour répondre au besoin de crédits des femmes se lançant dans l'exportation, permettrait aux femmes qui souhaitent exporter d'emprunter plus facilement.

Non seulement nous devons nous intéresser au développement des établissements dirigés par des femmes, mais nous devons aussi prendre en considération les centaines de milliers de femmes dans le monde qui essaient encore de pénétrer sur le marché de l'entrepreneuriat proprement dit. **[Nancy Carter, États-Unis, Accès aux marchés]**

Les réseaux de femmes ouvrent d'importantes perspectives au commerce international.

Les hommes chefs d'entreprise et décideurs ont depuis longtemps accès à des associations, des réseaux ou des clubs (d'affaires, ou de sport, comme les clubs de golf ou encore d'intérêt socioculturel) où ils discutent de leurs affaires. Les femmes entrepreneurs sont en général exclues de ces réseaux influents d'anciens élèves" ou des réseaux d'affaires informels bien établis et, de plus en plus, elles commencent à créer leurs propres réseaux. *Les Femmes Chefs d'Entreprises Mondiales* (FCEM) qui est un groupe de 34 pays dont le siège est aux Pays-Bas et la International Federation of Women Entrepreneurs (Fédération internationale des femmes chefs d'entreprise) dont le siège est à New Delhi sont les deux principaux groupes internationaux. Des milliers de groupes essentiellement locaux de femmes entrepreneurs existent dans des pays du monde entier. Ces réseaux, grâce à des contacts personnels et à la mise en commun du savoir-faire, ouvrent d'importantes perspectives aux femmes entrepreneurs.

"Sans coopération, il n'y a pas de progrès possible. L'entrepreneuriat "au féminin" est un facteur essentiel de la productivité économique. L'association *Les Femmes Chefs d'Entreprises Mondiales* (FCEM) est un organisme qui vise à encourager les femmes à devenir chefs d'entreprise et à exercer une influence dans ce domaine. Nous consacrons notre énergie à la mise en réseaux, à la formation, à l'encadrement et à l'éducation. Nos objectifs sont, notamment, les suivants : *i)* promouvoir énergiquement la formation des femmes à l'entrepreneuriat dans tous les pays de l'OCDE, en fournissant les informations nécessaires sur l'accès à l'entrepreneuriat et au travail indépendant ; *ii)* favoriser l'accès au crédit et au financement des entreprises dirigées par des femmes au démarrage et au cours de la phase initiale ; *iii)* améliorer les données et les statistiques sur les femmes entrepreneurs, particulièrement dans les PME. **[Joke van den Boer, Présidente de la FCEM, Pays-Bas]**

"En Inde, les femmes étaient en général peu enclines à prendre des risques. Mais aujourd'hui, la femme indienne chef d'entreprise se fait une place à part entière dans la société industrielle émergente. Le réseau *"International Federation of Women Entrepreneurs*" (IFWE) est une organisation récente – créée en 1993 – qui a fait connaître son existence dans le monde entier. **[Rajni Aggarwal, Secrétaire général de l'IFWE, Inde]**

Les pouvoirs publics commencent à encourager l'accès des femmes entrepreneurs aux marchés publics.

Les marchés du secteur public représentent un débouché important aussi bien pour les grandes entreprises que pour les PME, mais en raison de la discrimination et d'obstacles plus subtils, les entreprises dirigées par des femmes commencent seulement à avoir accès à ces possibilités. Dans certains pays, les pouvoirs publics favorisent activement l'accès des entreprises dirigées par des femmes aux marchés publics. Si au cours des deux dernières décennies, les États-Unis ont mis en place, du moins théoriquement, des politiques visant à attribuer aux entreprises dirigées par des femmes un plus grand nombre de marchés publics fédéraux, ces entreprises ne se voient encore attribuer qu'un petit pourcentage de tous ces marchés. Malgré cela, des progrès sensibles ont été réalisés ces dernières années suite à un effort concerté des décideurs fédéraux. Il semble que ces derniers s'emploient principalement à privilégier la participation des entreprises dirigées par des femmes, à normaliser les procédures et à encourager la formation.

Toutefois, les femmes peuvent se trouver exclues des possibilités offertes par le secteur des entreprises dans la mesure où elles ne font pas partie des "clubs masculins".

L'obtention de contrats de sous-traitance offerts par les grandes entreprises pose également un problème évident. Les grandes entreprises peuvent avoir des programmes dynamiques d'association avec des partenaires commerciaux, des fournisseurs et des sous-traitants, mais dans la mesure où les femmes entrepreneurs ne s'inscrivent pas en général dans les réseaux d'affaires influents – aux mains des hommes principalement – elles sont souvent exclues des possibilités offertes en la matière. Cela peut être vrai également des programmes de sous-traitance directe ou indirecte mis en place par les grandes sociétés multinationales qui ont encouragé les PME fournissant des biens et des services à commencer à exporter (par exemple, si un premier entrepreneur a vendu du matériel à une entreprise dans un pays X, un deuxième exportateur peut fournir des pièces de rechange et un troisième exportateur assurer une formation pour utiliser efficacement le matériel).

Conséquences pour l'action des pouvoirs publics et bonnes pratiques

Les politiques doivent veiller à ce que les femmes aient des possibilités de s'instruire, de s'informer et d'accéder aux marchés...

Dans de nombreux pays, les responsables et les décideurs se soucient de plus en plus d'aider les femmes à exporter et également de faciliter leur accès aux marchés intérieurs, aux plans national et local. Il faut donc que les femmes aient des possibilités de s'instruire, de s'informer et d'avoir accès aux marchés. Un certain nombre d'initiatives et de programmes gouvernementaux ou publics concluants ont été présentés durant la conférence.

Certains pays ont développé des programmes non sexistes et tenant compte du rôle spécifique des femmes pour aider les entreprises à exporter...

Par exemple, aux États-Unis, **David Bowie [États-Unis, Accès aux marchés]** a expliqué qu'il existait un grand nombre de programmes non sexistes pour aider les entreprises à exporter ou à accroître leurs exportations. L'initiative en faveur des femmes entrepreneurs s'est traduite essentiellement par une série de missions sur le thème "Les femmes et le développement des entreprises commerciales", qui offrent une excellente occasion aux participants d'établir des contacts d'affaires à l'étranger. L'initiative "Les femmes et le commerce" prévoit également d'informer les associations de femmes entrepreneurs de l'existence de programmes d'aide à l'exportation comme le US Trade Information Center et le Trade Base Programme (numéro d'appel gratuit, système d'information par Internet et fax pour aider les exportateurs ou importateurs potentiels qui peuvent recevoir de nombreux conseils sur les programmes fédéraux d'aide à l'exportation).

... et pour mieux faire connaître les programmes existants et favoriser l'accès à ces programmes.

Les autorités australiennes veillent également à ce que les femmes entrepreneurs aient connaissance de toute la gamme des programmes d'exportation en faveur des femmes et à ce qu'elles aient accès à ces programmes, comme l'a indiqué **Angeline Low [Australie, Mondialisation]**, qui a décrit comment AUSTRADE, l'organisme public chargé de promouvoir les activités commerciales, encourage les femmes entrepreneurs à exporter. AUSTRADE recense les femmes entrepreneurs qui jouent un rôle actif dans les exportations et qui peuvent servir de modèle à d'autres femmes et essaie de les aider à établir des réseaux avec des femmes d'autres pays afin d'améliorer les exportations. Un des problèmes majeurs, toutefois, est que les femmes sous-estiment en général les services publics et les services d'aide et de promotion du commerce.

Le but des réseaux électroniques est d'accroître la participation des femmes entrepreneurs aux échanges internationaux.

Les organisations internationales encouragent également ces tendances en mettant à profit les réseaux électroniques. Le réseau électronique de la CNUCED - les pôles commerciaux - a été créé afin de favoriser la participation au commerce international. **Christine Cattell [CNUCED, Mondialisation]** a expliqué que les pôles commerciaux étaient essentiellement des centres de facilitation des échanges où les négociants pouvaient obtenir des renseignements sur la manière d'effectuer des transactions internationales. Les pôles commerciaux servent également de points d'accès aux réseaux internationaux et lorsqu'il en existe dans divers pays, les organisations de femmes et les associations d'entreprises cherchent à participer au programme.

Faciliter l'accès au marché pour les entreprises dirigées par des femmes, signifie également leur garantir l'accès aux marchés publics.

Anthony DeLuca [États-Unis, Accès aux marchés] a exposé un programme novateur mis en place par la US Air Force pour faire en sorte que les femmes entrepreneurs aient accès aux marchés publics. M. DeLuca a expliqué que pour contribuer à ce que les entreprises dirigées par des femmes disposent effectivement d'informations sur les marchés publics en attente, la US Air Force avait adopté une règle pour que chacune de leurs activités d'achat mette en jeu au moins une entreprise dirigée par une femme pour chaque acquisition soumise à concurrence au-dessus du seuil d'acquisition simplifié (Simplified Acquisition Threshold - SAT). En outre, des listes d'entreprises dirigées par des femmes ont été données aux grandes entreprises pour qu'elles leur fassent part des possibilités de sous-traitance. En plus de l'aide qui leur a été apportée en matière de normalisation, les entreprises dirigées par des femmes ont eu accès aux activités de la US Air Force grâce à son tableau d'affichage électronique disponible sur Internet. La formation est un autre secteur important qui doit faire l'objet d'une attention suivie aussi bien pour les acheteurs de la USAF que pour les femmes entrepreneurs. Celles-ci ont été formées par la Air Force Business Education Team (Equipe de formation des chefs d'entreprise) aux modalités de gestion des entreprises et les acheteurs de la USAF ont reçu une formation pour leur permettre d'être mieux à même de traiter avec les femmes entrepreneurs. Ces efforts tendant à favoriser l'accès des femmes au marché de la USAF se sont soldés par une hausse de 25 pour cent sur deux ans.

Recommandations :

Les participants à la conférence se sont accordés à reconnaître qu'en matière d'exportations l'accès au financement et aux garanties était le moyen essentiel d'aider les femmes à développer leurs activités internationales. Toutefois, un certain nombre d'autres propositions ou recommandations ont été formulées pour encourager les femmes à exporter et à améliorer leur accès au marché intérieur. Il s'agit notamment :

◊ *D'assurer des programmes généraux d'enseignement/de formation. Par exemple, les conseillers commerciaux dans les ambassades devraient s'appliquer à comprendre les problèmes des femmes entrepreneurs et les obstacles auxquels elles sont confrontées lorsqu'elles veulent développer leurs affaires.*

◊ *De mettre en place des programmes de tutorat. Un programme international "tuteur-protégé" devrait être adopté pour aider les entreprises dirigées par des femmes, en mettant en commun et en édifiant les bases nécessaires à la croissance et en dispensant des conseils stratégiques.*

◊ *De recommander aux gouvernements d'alléger les charges bureaucratiques et administratives pour les PME dirigées par des femmes qui souhaitent exporter ou accéder à de nouveaux marchés.*

◊ *De relever le niveau de participation des femmes dans le cadre des marchés publics.*

◊ *Il faudrait encourager les pouvoirs publics à promouvoir les missions commerciales de femmes et leur participation à toutes les missions commerciales officielles.*

◊ *De favoriser la formation et le développement d'associations et de réseaux de femmes entrepreneurs afin de créer une infrastructure commerciale internationale par : des programmes éducatifs, un accès plus facile aux crédits, le recensement de marchés cibles, la mise au point de bases de données en ligne propres à recenser les possibilités de vente et d'achat, l'octroi de licences pour les systèmes d'automatisation en matière d'exportations et la sélection de partenaires fiables pour des coentreprises.*

◊ *D'encourager les alliances stratégiques/les coentreprises entre les femmes migrantes travaillant à l'étranger et les femmes se trouvant dans leur pays d'origine pour multiplier les débouchés.*

◊ *De mettre à profit le commerce électronique et de créer un marché mondial en vue d'établir un réseau technologique pour les entreprises dirigées par des femmes. Un système de communications et d'échanges d'informations d'accès libre devrait être appliqué concrètement et efficacement.*

◊ *De reconnaître la nécessité pour les organismes concernés de pouvoir faciliter les échanges internationaux d'experts entre pays, notamment en faveur des économies en développement et en transition, sur la question des femmes entrepreneurs.*

◊ *D'améliorer l'information sur les programmes d'aide en cours pour permettre aux petites entreprises d'avoir accès aux marchés et d'accroître le taux d'utilisation de ces programmes d'aide en s'employant plus efficacement à illustrer et à faire savoir comment et pourquoi les services profiteront aux femmes.*

◊ *De veiller à ce que des organisations internationales comme l'OCDE encouragent les gouvernements à échanger des renseignements sur les besoins des femmes entrepreneurs en matière d'accès aux marchés et à promouvoir la diffusion de pratiques exemplaires d'accès aux marchés parmi les entreprises dirigées par des femmes dans les pays de l'OCDE ainsi que dans les économies en transition et en développement.*

F. Transformer l'Environnement Réglementaire et Institutionnel

Principaux problèmes

Les femmes doivent exercer des pressions pour obtenir les moyens d'action nécessaires à leurs petites entreprises.

La conférence a axé son attention sur les activités des femmes entrepreneurs, mais elle a aussi clairement montré que les femmes voulaient pouvoir influencer la politique des pouvoirs publics de manière à ce qu'elle tienne compte de *leur* conception de l'économie, des principales orientations sociales et économiques pour l'avenir et de leurs besoins particuliers afin d'étayer cette conception. C'est pourquoi la plupart des ateliers ont examiné des questions liées aux cadres institutionnels et réglementaires propres à établir un environnement économique favorable aux femmes entrepreneurs.

Les petites entreprises passent beaucoup trop de temps à s'occuper de questions administratives.

Les PME se plaignent souvent des charges réglementaires et administratives qui pèsent sur leurs activités. **Jane Applegate [États-Unis, Gestion]** a fait observer que c'était là une des grandes difficultés auxquelles étaient confrontées les petites entreprises, lesquelles devaient consacrer beaucoup trop de leur temps à s'occuper de questions administratives et réglementaires, ce qui était très onéreux. Les grandes entreprises peuvent répartir cette charge sur un volume de production et un nombre de personnes beaucoup plus élevé, ce qui leur donne un avantage comparatif. Cela n'est guère conforme à la politique des pouvoirs publics qui vise à assurer des conditions égales à toutes les entreprises, indépendamment de leur taille. De nombreux pays ont adopté des mesures novatrices pour simplifier les réglementations, mais souvent le processus de réglementation proprement dit ne suit pas le rythme de l'évolution des économies et de la technologie.

Il semble que pour les femmes les charges réglementaires et administratives soient encore plus lourdes, dans la mesure où en général elles créent des petites et des micro-entreprises et où elles ont peut-être moins d'expérience pour acquérir des informations sur les réglementations administratives et effectuer un grand nombre de tâches administratives. **Naoko Bano [Japon, Gestion]** a illustré le problème de crédibilité des femmes entrepreneurs en exposant les difficultés auxquelles elle s'était heurtée en essayant d'obtenir auprès du service local compétent les renseignements nécessaires à l'approbation de produits.

Le statut juridique des femmes dans le monde des affaires devrait être clarifié.

Outre ces contraintes qui touchent toutes les PME, les femmes chefs d'entreprise rencontrent parfois d'autres difficultés liées à leur statut juridique. En effet, le choix d'une structure juridique pour une entreprise est un facteur déterminant qui peut avoir une incidence sur les niveaux de fiscalité, de rentabilité, de protection sociale et de pension de retraite. Les femmes co-entrepreneurs sont souvent tributaires de leurs partenaires pour ce type de prestations sociales et, dans de nombreux pays, le système fiscal est mal adapté à leur statut particulier. **Maria Teresa d'Avila [Portugal, Financement]** a indiqué que les organes de recours n'étaient pas en mesure de régler ces différends car la loi n'abordait pas explicitement la question. C'est pourquoi dans une entreprise familiale, la reconnaissance légale du rôle et des droits des femmes par rapport aux autres membres de la famille est fondamentale.

Un cadre réglementaire souple et efficace : une condition préalable.

Les participants ont souligné la nécessité d'un cadre réglementaire plus souple et plus efficace pour les très petites ou les micro-entreprises qui sont représentatives des entreprises dirigées par des femmes. Un cadre réglementaire plus souple et plus efficace, de même que des coûts de main-d'oeuvre non salariaux moins élevés, aideraient également les femmes entrepreneurs durant la phase de démarrage, comme l'a fait observer **Florence Lardenchet [France, Dialogue...].** Dans les économies en transition, la rigidité de la réglementation et l'absence de structures commerciales appropriées constituent de sérieux obstacles pour les femmes entrepreneurs et les très petites entreprises. Selon **Josianne Capt [OIT, Dialogue...]**, dans les pays en développement, la majorité des micro-entreprises dirigées par des femmes ont peu de chances de s'agrandir ; nombre d'entre elles appartiennent au secteur informel et ne tiennent pas compte de la plupart des normes internationales élémentaires du travail (et ce, alors que dans de nombreux pays en développement les femmes ne puissent posséder de terres et que leur accès aux moyens de production soit limité). La question est de savoir si oui ou non ces micro-entreprises peuvent se développer au point de devenir des entreprises officielles qui respectent les normes du travail et offrent des prestations sociales.

Des structures d'appui plus efficaces sont nécessaires pour les entreprises dirigées par des femmes. **Lois Stevenson [Canada, Dialogue...]** a fait observer que les femmes avaient dans une large mesure elles-mêmes répondu à ce besoin en formant des associations pour accéder plus facilement aux marchés et pour transformer l'environnement réglementaire et institutionnel. Ces associations fournissent également une aide, des conseils et une formation et catalysent les possibilités offertes aux femmes entrepreneurs de tisser des réseaux avec les responsables des marchés publics et privés.

"Dans de nombreux pays, les femmes entrepreneurs n'utilisent pas les ressources ou les réseaux existants. Comment les pouvoirs publics pourraient agir plus efficacement, de concert avec les associations de femmes entrepreneurs et d'autres organismes, pour faire en sorte que les femmes ne réinventent pas la roue, mais pour l'essentiel achètent et mettent à profit les ressources qui existent déjà". **[Virginia Littlejohn, États-Unis, Mondialisation]**

Les femmes considèrent qu'il est nécessaire d'élaborer des politiques et des initiatives, de les contrôler et de les soutenir activement. Leurs associations et leurs réseaux leur permettent d'approfondir leur connaissance des processus et des méthodes de sensibilisation permettant d'influencer le législateur, qu'il s'agisse de campagnes de sensibilisation auprès du pouvoir exécutif et législatif à l'échelon local ou au niveau des États et de la nation, ou d'activités efficaces de mobilisation au plan multilatéral et international.

"Si les femmes contribuent effectivement à la vie économique – en payant des impôts, en utilisant le crédit et en créant des emplois et des ressources monétaires – alors les pouvoirs publics doivent entendre notre voix. Comment pouvons-nous légitimer l'esprit d'initiative dont les femmes peuvent faire preuve et nous asseoir à la table où se trouvent certains des principaux décideurs, notamment les représentants des banques centrales, les PDG ou les hommes politiques ? Les femmes chefs de PME sont-elles à la table du Forum économique mondial ? A la Chambre internationale de commerce ? Dans les conseils des multinationales ? Car c'est là que réside en grande partie le pouvoir de décision. Existe-t-il d'autres organisations que nous pourrions créer pour aider à définir le programme d'action mondiale ? Les femmes chefs de file symbolisent souvent le changement d'une manière que l'on ne retrouve jamais parmi ceux qui exercent effectivement le pouvoir. **[Laura Liswood, États-Unis, Première séance plénière]**

Conséquences pour l'action des pouvoirs publics et bonnes pratiques

Grâce aux associations, les femmes commencent maintenant à influer sur le processus de décision.

Dans certains pays, les femmes ont franchi d'importantes étapes dans les domaines examinés ci-dessus. Aux États-Unis, le National Women's Business Council, présenté par **Amy Millman [États-Unis, Environnement économique et technologie]**, a été créé pour donner à des groupements comme la National Association of Women Business Owners (NAWBO) et 700 autres associations d'entreprises dirigées par des femmes, la possibilité d'influer sur le processus de décision. Cet organisme veille à ce que les initiatives publiques tiennent compte de leurs priorités et que d'autres établissements du secteur privé aient connaissance de leurs contributions. Un forum a réuni, d'une part la National Association of Women Business Council et diverses associations de femmes et, d'autre part, des organismes bancaires et financiers et des établissements d'enseignement afin de mettre en place un programme induit et conçu par les femmes qui illustre l'économie telle qu'elles l'imaginent au cours de la prochaine décennie et aussi telle qu'elles la souhaitent, et les infrastructures d'appui nécessaire à cette fin.

Les promoteurs de cette initiative travaillent en étroite collaboration avec l'Interagency Committee on Women Business Enterprise des États-Unis (Comité inter-organisations sur les entreprises dirigées par des femmes) qui se compose notamment de hauts fonctionnaires de grands ministères et organismes du gouvernement fédéral. L'objectif est d'étudier comment traduire les besoins des Américaines en actions concrètes, en politiques et en programmes. Les étapes suivantes consistent à mettre au point des réseaux locaux permettant de diffuser l'information et de faire connaître aux femmes les possibilités que le gouvernement fédéral leur offre, ainsi que des renseignements propres à favoriser la croissance de leur entreprise. Au cours des deux prochaines années, les travaux porteront sur des questions liées au développement des entreprises dirigées par des femmes.

Les programmes de formation à la gestion gagnent du terrain...

Susan Davis [États-Unis, Dialogue...] a indiqué que le Center for International Private Enterprise (CIPE – Centre pour l'entreprise privée internationale), qui fait partie de la Chambre de commerce des États-Unis, avait également élaboré à l'intention des femmes des programmes de formation concernant les associations professionnelles. Au cours des dix dernières années, la participation totale de femmes à ces stages de formation a fait un bond, passant de 7 pour cent en 1986 à 32 pour cent en 1995. En dehors de ce type de formation, le CIPE offre également des aides directes à des organismes pour les aider à se développer. Son objectif est de promouvoir le plus possible l'égalité des chances en faveur des femmes entrepreneurs, en donnant effectivement davantage de pouvoir aux associations féminines. En effet, pour le CIPE, les femmes entrepreneurs peuvent, et doivent, définir la politique gouvernementale par l'intermédiaire des associations professionnelles et par des activités de mobilisation de l'administration publique, le développement des capacités, la gestion des affaires publiques, l'utilisation des médias, la gestion financière, etc.

... et, de plus en plus, les femmes participent aux débats de haut niveau.

L'Australian Council of Business Women (ACOB – Conseil australien des femmes d'affaires) a été créé en novembre 1995 pour faire en sorte que les opinions, les idées et les fonctions des femmes d'affaires australiennes soient prises en compte dans les débats de haut niveau et pour influer sur la création de richesses et d'emplois. Le Conseil s'emploie à définir comment élaborer, contrôler et soutenir activement des politiques et des initiatives, à rassembler et à diffuser des informations et des données de la recherche pertinentes et à favoriser le succès des entreprises dirigées par des femmes. Il veille également à définir des modalités de participation aux réseaux internationaux **[Angeline Low, Australie, Mondialisation].**

Dans certains pays, les pouvoirs publics ont pris des initiatives pour aider les femmes entrepreneurs.

Uma Reddy [Inde, Dialogue...] a signalé que le gouvernement indien avait pris un certain nombre d'initiatives pour aider les femmes entrepreneurs, notamment sous la forme de mesures financières. Les femmes entrepreneurs se sont également organisées en associations pour s'entraider. AWAKE ou Association of Women Entrepreneurs (Association des femmes entrepreneurs) de Karnataka fait appel à l'action bénévole de femmes d'affaires ayant réussi. Elle assure une aide en matière de formation, de gestion des entreprises et de conseil et offre une pépinière d'entreprises. AWAKE dispose de son propre centre de recherches et de documentation et applique un programme de renforcement des capacités de gestion de nature à favoriser la croissance des entreprises dirigées par des femmes. Elle est également affiliée à la Banque mondiale des femmes. Elle a obtenu de bons résultats et le gouvernement lui a demandé d'entreprendre un programme industriel rural.

Recommandations :

Les participants aux ateliers de la conférence ont fait de nombreuses recommandations pratiques visant à transformer l'environnement réglementaire, institutionnel et économique. En particulier, ils ont concentré leur attention sur l'aptitude des femmes à influencer la politique générale des pouvoirs publics et les structures, et envisagé les moyens d'aider les PME en général et les entreprises dirigées par des femmes, en particulier :

◊ *La culture de l'entreprise devrait être promue en vue de développer la capacité des individus à saisir des opportunités d'affaires, à prendre des risques et à innover.*

◊ *Comment les pouvoirs publics peuvent-ils régulièrement réviser la législation pour faire en sorte que les politiques n'aient pas tendance à favoriser les grandes entreprises, ce qui pourrait limiter la concurrence en éliminant les petites entreprises et les concurrents indépendants ? Le rôle de la politique financière dans l'aide apportée aux femmes entrepreneurs a également été souligné.*

◊ *Encourager et définir des fonctions de sensibilisation pour les organismes intéressés en vue d'élaborer des lois, de limiter les réglementations qui ont une incidence sur les PME et de créer des possibilités d'accès aux marchés publics dans les secteurs privé et public.*

◊ *Les femmes chefs d'entreprise doivent intervenir pour réduire les charges administratives dans le secteur privé, étant entendu que dans de nombreux pays les administrations travaillent déjà de concert avec le secteur privé pour délimiter et réduire le poids des réglementations.*

◊ *Les femmes chefs d'entreprise, en particulier celles qui travaillent dans de petites entreprises ou à domicile et qui risquent de se sentir isolées, devraient mieux connaître les réseaux et ressources existant en matière de formation et de savoir-faire, diffusion des technologies, établissement de contacts de nature à faciliter l'accès aux marchés, etc. et pouvoir s'intégrer plus facilement à ces réseaux et ressources.*

◊ *Les associations et réseaux de femmes devraient être renforcés, en particulier en mettant en relation les associations locales et nationales par l'intermédiaire des réseaux internationaux, de manière à ce que les femmes puissent collectivement accéder aux ressources économiques et aux moyens d'action.*

◊ *En tant qu'organisation internationale, l'OCDE devrait encourager les pays Membres à prendre en compte les femmes entrepreneurs dans l'élaboration de la politique gouvernementale concernant les PME, notamment dans l'élaboration des mesures visant la coopération pour le développement.*

◊ *Les organisations internationales comme l'OCDE pourraient contribuer à l'intégration des questions relatives aux femmes dans les pays non membres ainsi que dans les économies en développement et en transition, et à en faire un thème central des programmes de développement. Elles pourraient également favoriser le dialogue sur l'action à mener en la matière entre les pays bénéficiaires et les pays Membres.*

Conclusions

En partenariat avec les hommes, les femmes doivent investir dans les ressources humaines, la technologie et la création de connaissances, ainsi que dans la mise en place des nouvelles structures nécessaires à un développement économique général.

La conférence a amorcé un dialogue entre décideurs, chefs d'entreprise et institutions sur les femmes et la création d'entreprises et de richesses. A travers cette conférence l'OCDE a contribué au développement du potentiel de l'entrepreneuriat féminin. Les femmes entrepreneurs ont clairement indiqué qu'elles ne dépendaient pas seulement des gouvernements pour atteindre ce but. En partenariat avec les hommes, elles doivent investir dans les ressources humaines, la technologie et la création de connaissances, ainsi que dans la mise en place des nouvelles structures nécessaires à un développement économique général. Il faut établir des points de liaison entre les femmes et les hommes dans l'entreprise, entre pays et entre pays développés et en développement.

Aujourd'hui, nous parlons des femmes et de la création d'entreprises et de richesses. Ce dialogue touche à l'essentiel de la politique économique telle que la conçoit l'OCDE et traduit notre conception des sociétés économiques et du rôle important que les femmes doivent jouer pour optimiser les talents et les possibilités afin que nos économies restent à l'avenir compétitives. **[Chris Brooks, OCDE (TDS), Séance plénière de clôture]**

Les femmes doivent avoir une vision de l'avenir...

Les participants à la conférence ont reconnu la nécessité d'adopter une vision prospective et appuyé les efforts tendant à définir un programme de changement. "Nous devons envisager l'an 2000 comme point de départ pour décider de ce que nous voulons promouvoir, compte tenu des recommandations pratiques que cette conférence a permis de dégager". **[Ronnie Feit, États-Unis, Première séance plénière]**

... ainsi qu'un plan d'action et un mécanisme de contrôle.

Certaines recommandations doivent être appliquées au niveau des gouvernements et à l'échelon multilatéral, mais le secteur privé - banques privées, entreprises et associations de femmes entrepreneurs - ont également un rôle à jouer dans la suite à donner à ces recommandations. "Il nous faut non seulement une vision commune ou une stratégie d'exécution, mais aussi une ferme détermination, un plan d'action, un calendrier qui déterminera quand nous devons agir, et un mécanisme de contrôle pour ceux qui suivront l'action entreprise". **[Virginia Littlejohn, États-Unis, Mondialisation]**

"Si les entreprises dirigées par des femmes continuent à se développer comme elles l'ont fait jusqu'à présent, nous pouvons imaginer que nous posséderons au moins la moitié des avoirs dans le monde. Un proverbe chinois dit que les femmes soutiennent la moitié du ciel. Nous pourrons plus sûrement soutenir notre propre moitié si nous possédons le terrain sur lequel nous nous tenons." **[Ronnie Feit, États-Unis, Première séance plénière]**

Recommandations de la Conférence

Les principales recommandations qui se sont dégagées de la Conférence sont énoncées ci-après :

◊ *Mieux appréhender la contribution des femmes entrepreneurs à la croissance économique, aux échanges internationaux et à la création d'emplois dans les pays de l'OCDE en transition et en développement, en améliorant les données et les statistiques relatives aux PME appartenant à des femmes. De même, améliorer l'harmonisation statistique entre les pays de façon à faciliter l'analyse comparative.*

◊ *Mieux comprendre les obstacles que rencontrent les entreprises appartenant à des femmes et dans quelle mesure la profession et les décideurs politiques réussissent à s'attaquer au problème.*

◊ *Encourager des mesures gouvernementales qui instaurent des règles du jeu équitables et améliorer la compétitivité des entreprises dirigées par des femmes, par exemple en prenant des dispositions pour supprimer les obstacles législatifs et réglementaires, sources de distorsions.*

◊ *Supprimer les difficultés excessives que rencontrent les femmes entrepreneurs pour se procurer des capitaux et mettre au point des instruments financiers qui répondent à des besoins en capitaux légitimes à tous les stades de développement des entreprises (micro-entreprises, petites, moyennes ou grandes entreprises). Apporter une attention spéciale aux besoins financiers des entreprises dans le secteur des services. La recherche devrait s'attacher à comparer l'expérience d'un pays à l'autre entre économies de l'OCDE, en développement et en transition, et analyser la nature et la valeur du "capital intellectuel" de façon à donner aux établissements financiers de nouveaux éléments pour évaluer le risque-crédit.*

◊ *Réserver un accueil favorable aux initiatives privées en faveur de la création d'un consortium international de banques visant à renforcer la compétence du secteur financier en matière de financement des entreprises dirigées par des femmes et à le sensibiliser aux problèmes spécifiques qu'elles rencontrent.*

◊ *Encourager les organismes d'aide et les banques régionales de développement multilatérales à mettre au point des programmes permettant aux femmes entrepreneurs d'avoir accès aux capitaux dans des conditions d'égalité ; l'égalité des chances devrait aussi s'appliquer aux programmes relatifs aux données et statistiques, à la gestion et à l'assistance technique, à la constitution d'associations, à l'utilisation des technologies, à l'accès aux marchés et aux échanges internationaux.*

◊ *Mettre au point des méthodes d'analyse référentielle de nature à rendre compte de la contribution unique des entreprises dirigées par des femmes partout dans le monde et à répondre à leurs besoins spécifiques. En ce qui concerne le financement, l'analyse référentielle pourrait porter sur les possibilités données aux "investisseurs informels" (business angels), sur la formation de capital sous forme de fonds propres ou de quasi-fonds propres et sur l'élaboration de programmes créatifs de garanties assortis d'une formation appropriée.*

◊ *Recenser partout dans le monde les "meilleures pratiques" relatives à l'entreprise "au féminin" et en faire la synthèse via les associations et les ONG, Internet, et aussi les gouvernements et les organismes multilatéraux. L'OCDE devrait apporter son concours à la centralisation et à la diffusion de l'information sur les meilleures pratiques recensées.*

◊ *Encourager le système éducatif, à tous les niveaux, à dispenser une formation à la gestion et à fournir des connaissances de base dans le domaine de l'économie, sans discrimination entre les deux sexes, tout au long du processus éducatif à partir, notamment, des "meilleures pratiques" recensées et d'études de cas relatant des expériences réussies.*

Recommandations de la Conférence *(suite)*

- *Encourager au sein des économies de l'OCDE, en développement et en transition, l'adoption d'initiatives accessibles et d'un coût abordable dans les domaines de l'enseignement et de la formation afin d'aider les femmes entrepreneurs à acquérir des compétences en matière de gestion, à accéder aux marchés et à se doter d'un savoir-faire technologique pour leur permettre de faire progresser rapidement leur entreprise. La formation et l'enseignement doivent être suffisamment souples pour s'adapter à la situation locale particulière et répondre aux besoins de formation à tous les stades de développement de l'entreprise.*
- *Encourager la formation et le développement d'associations et de réseaux de femmes entrepreneurs à des fins de formation et d'acquisition de savoir-faire, de diffusion des technologies, de façon à établir des contacts qui facilitent l'accès aux marchés, etc. Les associations nationales doivent être encouragées à mieux faire circuler l'information auprès des femmes entrepreneurs, et faciliter la nomination de femmes à des postes de conseillers des gouvernements et à des postes de décision. En outre, encourager les femmes entrepreneurs à participer aux activités des institutions économiques en place comme les Chambres de commerce.*
- *Mettre au point des mécanismes visant à faciliter l'accès aux marchés, y compris en ce qui concerne les méthodes de passation de marchés, publics ou privés.*
- *Mettre au point des initiatives visant à faciliter l'accès aux marchés internationaux, au travers notamment de missions commerciales, des liens établis entre associations de femmes entrepreneurs au plan international, du commerce électronique, etc. Les gouvernements devraient rendre plus facile l'accès au financement pour les femmes entrepreneurs dont l'activité porte sur les échanges internationaux (financement des exportations-importations, y compris assurance sous forme de garantie des exportations).*
- *Elaborer un calendrier de recherche sur les femmes entrepreneurs et encourager experts, gouvernements, organismes de développement, organisations multilatérales et associations à s'appliquer à "combler les lacunes". L'OCDE doit jouer un rôle important afin de mobiliser les énergies pour définir l'action à entreprendre et de veiller à ce que ce calendrier soit effectivement mis en oeuvre.*
- *Organiser, dans le prolongement de la première, une deuxième conférence des femmes entrepreneurs à la tête de PME, à l'OCDE, en l'an 2000, afin d'évaluer les progrès réalisés dans la mise en oeuvre des recommandations formulées en 1997, analyser l'environnement économique dans lequel évolueront les entreprises et faire de nouvelles recommandations pour les cinq années suivantes.*

Notes

1. Extrait de la présentation de Mme Shigeko Mitsuhashi à l'atelier 6 sur l'accès aux marchés. Le tableau lui-même est extrait d'un manuel destiné aux femmes créateurs d'entreprises du Nikkei Research Institute of Industry and Markets.

2. Conférence intergouvernementale de l'OCDE sur "l'éducation et l'économie dans une société en mutation", 1988 et le rapport conjoint OCDE-CERI (1989), *Apprendre à entreprendre : nouvel enjeu pour l'éducation et la formation*, qui débat des questions fondamentales liées aux possibilités et aux obstacles résultant de l'introduction de l'esprit d'entreprise dans l'éducation et la formation.

3. Le Centre de recherche sur l'éducation et l'innovation de l'OCDE a entrepris un certain nombre d'activités sur la formation continue et l'investissement en capital humain. Voir aussi le rapport de l'OCDE (1998), *Investir dans le capital humain : Une comparaison internationale.*

Bibliographie

BARRETT, Mary (1997), Rapport régional sur l'Asie et le Pacifique, établi sous forme de document de travail non publié pour la conférence de l'OCDE.

BRUSH, Candida (1990), "Les femmes et la création d'entreprises", dans Sara K. Gould et Julia Parzen (éds), *Entreprendre au féminin. Initiatives locales de création d'emplois,* OECD/ILE, p. 39.

CATTAN, Claude (1997), Rapport régional sur l'Italie et l'Europe méridionale, établi sous forme de document de travail non publié pour la conférence de l'OCDE.

DUCHENEAUT, Bertrand (1997), "Women Entrepreneurs in SMES", Rapport rédigé pour la conférence de l'OCDE.

HOLMQUIST, Karin (1996), "Female Enterprises: Current Situation and Development Trends" dans *Nordic Labour Journal 2/1996.*

MARLEAU, Martine (1994), "L'Entrepreneurschip au féminin", rapport de la *"Fédération des Entreprises Indépendantes",* décembre.

NFWBO (1997), *1996 Facts on Women-owned Businesses: State Trends,* The National Foundation for Women Business Owners, Silver Spring, MD, États-Unis.

OCDE (1994), *Les femmes et le changement structurel : nouvelles perspectives,* Paris.

OCDE (1996*a*), *Technologie, productivité et création d'emplois,* Vol. I, Synthèse et conclusions, Paris.

OCDE (1996*b*), Atelier de haut niveau sur les PME : *l'emploi, l'innovation et la croisance,* Paris.

OCDE (1997*a*), *Statistiques de la population active,* Paris.

OCDE (1997*b*), *PME et mondialisation,* Vol. 2, études par pays, Paris.

RUNEBORG, Silvia (1997), Rapport régional sur la Suède et l'Europe septentrionale, établi sous forme de document de travail non publié pour la conférence de l'OCDE.

SKINNER, Jane et Jane SLOWEY (1996), Rapport pour l'OCDE, Programme LEED, octobre.

SMALL BUSINESS ADMINISTRATION (1997), *Statistics on Women Business Ownership.*

STARCHER, Diane Chamberlain (1996), "Femmes entrepreneurs : Catalyseurs de transformation", Association européenne baha'ie de management.

WEEKS, Julie (1997), Rapport régional sur l'Amérique du Nord et quelques pays d'Amérique latine, établi sous forme de document de travail non publié pour la conférence de l'OCDE.

WEISS, Chris (1990), "Le rôle joué par les organismes intermédiaires pour renforcer le travail indépendant des femmes" dans *Entreprendre au féminin,* OCDE/ILE, Paris.

Part II:

Selected Papers

Chapter 2

WOMEN ENTREPRENEURS IN SMEs: A VISION FOR THE FUTURE

by
Rona F. Feit
Rona Feit and Associates, United States

In the new digital economy, women, like men, will create wealth by applying knowledge, networked human intelligence and effort to whatever business sector they choose to be in – services, retaining, manufacturing and even agriculture. They will have new access to sources of capital through new computerised investor networks. They will be able to tap the talents of people all over the world. They will be able to do business everywhere. In many important ways they will be able to do more with less and do it more quickly.

In the old economy, being big helped to make a company successful. Today, being big is often a liability. Innovation, agility and organisational learning are the key variables for success. Bureaucratic hierarchies are out. The best new enterprises will probably be networks of distributed teams that act as clients and servers for each other.

This paper is an attempt to throw light on how a changing world might effect the potential of women's businesses to contribute to the economic health of the advanced economies of the world. It addresses two major challenges to that potential: a globalising economy that is shaking up established economic patterns, social structures and governments; and an information and telecommunication revolution underlying the global economy that empowers good and bad alike. It also considers what might be done by women business owners themselves to meet those challenges.

The challenge of the global economy

The invention of the computer and the Internet has sparked a world-wide explosive cycle of renewal, migration and destruction. These new tools have enabled the expansion of new kinds of industrialisation to new countries, some that previously had mainly agricultural or resource-extraction economies, and many that had been very poor and out of the main stream of modernity. Their lower costs and the capabilities of their people are attracting businesses from developed countries to relocate many of their operations there. This industrialisation has brought to these countries new opportunities, the learning of new skills, broader outlooks, new relationships, the accumulation of wealth and the rise of a new commercial class. It has also brought social disruption, exploitation of the weak (especially women), pollution, official and private corruption, and competitive practice that is accountable only to the bottom line.

Some have described the global economy created by the production and sale of goods and services world-wide as a competitive free-for-all. Companies feel they must relocate to any country that offers a chance to operate at lower cost or get some other competitive advantage such as access to a bigger market. Countries hungry for foreign investment give them what they want. These companies are producing a constantly changing array of products to attract consumers – some say a glut of goods. Most of the goods produced in the emerging economies are sold in advanced economies, but the producing nations often cannot afford to buy a similar amount of goods from the advanced economies – hence the negative balance of trade experienced by some developed nations, including the United States. Because of job losses and wage depression in advanced economies, the worry is that there will eventually not be enough buyers for all this production – the global economic bubble could burst.

Many leaders and thinkers take a more sanguine view. They see this global competition as wild but manageable. They believe it is a process that will spread the world's wealth more evenly, eventually creating a middle class everywhere which will demand justice, freedom and democracy. They think the developed world can compete successfully without sacrificing its democratic values or community life, and that the developing nations will sooner or later provide vast new markets for the goods produced by companies owned by citizens of developed nations.

The critics are not impressed by these views. They see a dangerous erosion of civilised standards in both the developed and the developing world. They note that the profits of enterprise are increasingly going to managing elites and investors with no national or community loyalty, while laid-off middle managers and workers at home suffer losses in income and job security. They fear the rule of market-place morality and rising individualism. They do not believe that an open market-place is a substitute for a social strategy. They doubt the ability of the advanced economies to keep investment and jobs at home or to craft international agreements that will preserve standards of fair labour practice or environmental protection. They point out the current downward trends in developed countries in economic growth, joblessness and pay scales for the many, that support their fears.

With either view, the arrival of the global economy is clearly not an unmixed blessing. It poses political questions that have not yet been answered satisfactorily, and there is an increasing call by citizens' groups for corporate accountability to civilised standards and to the nations and people who have chartered them. What can the global economy mean to women entrepreneurs? It offers great opportunities to those who are positioned for growth and have, or can develop, the right networks. But doing business in it may challenge even the most cosmopolitan women in terms of their values as well as their management of risk. The global economy may also be a threat to many of the smaller local women's enterprises with few or no employees that make up the bulk of women's enterprises.

There are many changes to which women entrepreneurs will have to respond. Let us consider just three: the power shift that the new global economic competition has caused in the world; the threat to the economic and social base of the communities in which women live and work; and the changing environment for the very small businesses that most women own.

A variety of commentators have noted that the globalising economy has led to a world-wide power shift. This shift is fundamentally a loss of influence and autonomy by nation states, both democratic and autocratic, who find that their national standards and rules can be avoided by the players in the international market-place. The impact of this problem and other changes have opened governments to the need for sharing power with business and the non-governmental citizens' groups (NGOs) that make up civil society. Governments' view of citizens is also changing: they are beginning to see them as customers they have to please – and who are presently quite disenchanted with them – and as agents of change and development whose help they must have to solve social and economic problems.

In this new policy environment, the ideas, energies and power of women are having increasing impact. This has been particularly evident at the United Nations. Women's NGOs in great numbers have been advocating for policies that empower women socially, politically and economically at UN conferences on the environment, human rights, population, social concerns, the human habitat, and women. They have also been among the voices calling for multinational corporations to be accountable to international standards, and for the reform of the international Monetary Fund and the World Bank. They have understood that the global economy has moved problems up to the international level, and that they need to be working at that level. They have successfully pressured to be, as one observer puts it, "out of the hallway" and "around the table".

The reasons for this shift in power are several. Key causes include the end of the Cold War, the apparent triumph of free-market capitalism and the impact of the new information and communication technology. Another is the fact that nations can no longer hide what they are doing from international scrutiny or keep their people uninformed about the world. International public opinion can now force previously insulated elites to respond to outside influence. The shift also reflects the lessened threat of global war. The nuclear stalemate, international disarmament efforts and spy satellites have made military might less useful and alarming. The global economy is now dominated by multinationals, private investors and currency traders with few national ties and loyalties only to the bottom line. Governments, in fear of losing footloose multinationals or experiencing a flight of investment capital that devalues their bonds and their currencies, mostly cater to market forces and are cautious about trying to control them.

The power shift has a significant bad side which derives directly from the information and telecommunication revolution. Private interests that are not positive, such as organised crime, extremist groups and the drug trade, already have little to fear from government. The era of networked intelligence works to their advantage – and a solution is not yet in sight. Using the Internet and unbreakable encryption they can organise, transfer and hide cybercash, and elude control. They can invade confidential records stored on the information highway, and steal or destroy intellectual property or engage in blackmail based on this ability.

This is a major problem for public order, democracy and business in general. Women entrepreneurs, like others, will have to be concerned about protecting the privacy of digital records and communications, and about knowing who they are dealing with.

The global economy will have an increasing impact on the health of local communities. Cities in developed countries will continue to attract people from declining rural areas and immigrants from everywhere.

Some communities will become crowded and have no jobs: others will empty out because there are no jobs. The ideal of "sustainable development" will continue to be a challenge in both urban and rural settings. As companies move around the world seeking competitive advantage in low-cost labour, local skills or entry to desired markets, they uproot the employment and social structure of the communities they leave. Joblessness in the abandoned communities increases dramatically. Even if businesses remain, they tend to downsize in order to lower labour costs.

Job security is becoming a thing of the past, and the new concept of "employability security" really means that every employee has to be highly marketable, mobile and entrepreneurial on his or her own behalf. While downsizing can lead to outsourcing opportunities for local small businesses, these may be offset by increased competition from companies based elsewhere. Given these conditions, plus new economies of scale and scope in retailing and services and the restructuring of wholesaling due to new technologies – all areas where women's businesses are concentrated – small local firms in affected communities are disappearing.

In this new world in which place and proximity are less and less important to business, women entrepreneurs concerned about the development of their children, the care of their ageing parents and the nature of home, may find themselves living in deteriorating towns and cities that do not support family, civic or business life. This is not news: "hollowed out" communities are already here. Government at all levels is already suffering an erosion of its tax base and becoming increasingly unable to make the community investments necessary to stem the downward spiral.

One of the few commentators who does not believe that community destruction is a necessary consequence of the global economy is Rosabeth Moss Kanter, a well-known professor at the Harvard Business School. In her recent book, *World Class*, she notes that a new wave of social concerns and a growing desire for community spirit have come in the wake of economic change. She asks if there is a basic conflict between social and community interests that are largely domestic or even local, and business competitiveness issues that are often international in scope. Unlike many other authors, her answer is to this question is "no". She thinks the global challenge can be managed to the advantage of both businesses and communities, and that "cosmopolitans" and "locals" can be united around a vision of excellence and expanded choices.

Kanter retains a focus on the importance of communities to economic growth that echoes in some respects the thinking of another woman economic theorist, Jane Jacobs. Jacobs' 1980's book, *The Wealth of* Cities, made the influential point that "the wealth of nations" has always depended on the existence of metropolitan centres in which businesses, over time, developed clusters of suppliers, innovators, traders and "know-how" that improved quality, designed new products and services and ways of producing them, opened new markets and created jobs and wealth. Kanter does not clearly say that such metropolitan centres are still necessary for the creation of wealth, but she is concerned to make them economically relevant in order to promote social cohesion and head off community destruction. She calls for cities to develop "core capability" in order to attract a mix of companies that are "thinkers", "makers" and "traders," to become centres of globally relevant skills with a central orientation that is flexible and can be renewed, updated and generalised across this mix. She stresses that cities, states and regional economies will only flourish to the extent that they provide linkage to global activities and networks.

The future of the small businesses she calls "locals" will depend on whether or not the communities in which they are located build an "infrastructure for collaboration" that breaks down all kinds of divisions between people, neighbourhoods and organisations, and works actively for economic development and civic improvement. The problems vary from place to place, but often include public cynicism, mutual distrust between cosmopolitan knowledge workers and less educated blue-collar and personal service workers, the politics of opposition, poor public education and a generally negative business climate. Kanter calls for both cities and companies to become "world-class". Being world-class means having the three Cs: *concepts* – the best and latest knowledge and ideas; *competence* – the ability to operate at the highest standards of any place anywhere; and *connections* – the best relationships, providing access to the resources of other people and organisations around the world. Cities which are world-class and collaborate for civic improvement are the best hope for women's very small businesses.

Women business owners will need to consider their responsibility in creating infrastructures of collaboration in their own communities. Kanter stresses that strong and inspired leadership is required. As encouragement, she gives examples of cities that have successfully remade themselves into world-class competitors. Her's is a hopeful approach, but it requires long and hard work, and success is not guaranteed. Women entrepreneurs, acting separately and together, using networked intelligence to make the job manageable, could take the load and make a difference – but they will need help.

The question of the global economy's impact on the different sizes and kinds of women's businesses will need much thought. The larger women's businesses with paid employees clearly have the potential to profit from the expansion of markets and the new suppliers of goods and services the global market-place can provide. In the United States, these businesses make up only 20 per cent of women's businesses but they account for 94 per cent of the dramatic growth in their receipts. They will need help with export/import procedures and with building new contacts and alliances, but they should be able to compete if they decide to do so. But, what about the other 80 per cent?

These very small businesses in the United States in 1992 accounted for most of the huge growth in the number of women-owned businesses. While they can be seen as a seedbed for future growth, there has been a lack of longitudinal research to tell us which ones will grow and why. We know that most of them have not grown much in the past. Many of these businesses are probably "income substitution" firms formed to create income where jobs have been lost or wages have been cut. Many of them are part-time or sideline businesses.

This lop-sided picture is a matter for concern. Many have worried for some time whether we are seeing the birth of a two-tiered enterprise economy as well as a two-tiered employment economy – with few bridges from one tier to the other in either. The question becomes even more urgent as developed nations embark on cutting or ending welfare payments and begin to demand that everyone who is not disabled or excused for other socially approved reasons, work for a living.

In the United States, since the late 1980s, citizen activists have developed a domestic microenterprise development movement that is trying to make enterprise a route to economic empowerment for non-traditional entrepreneurs. While it has drawn inspiration from the international microenterprise movement working in the developing world, it has learned that different economic and social conditions in advanced economies require different programme designs. It now includes hundreds of new, small, non-profit programmes around the country. They may provide small amounts of credit (ranging typically from $500 to $25 000) or business training, counselling and technical assistance, or both, to non-traditional entrepreneurs. They are financed by a combination of public and private funds, some from banks who are required under US law to show a record of investment in the communities from which they draw deposits.

Another level of alternative business and community development programmes also seeks to address the enterprise and community needs of our poor and low-income people. Leading actors include non-profit Community Development Finance Institutions (CDFIs) and non-profit Community Development Corporations. These organisations provide funding and leadership for citizen efforts to build and rehabilitate housing, develop businesses and improve social services. Their efforts have been crucial to filling the financing gaps created as traditional lenders consolidate, automate and move out of local communities, and to meeting development needs in the poor and low-income communities which federally funded development projects have generally failed to address.

There have been many success stories showing that, with their help, their low-income and poor clients, a majority of whom tend to be women, can build income and assets through enterprise. This has led slowly to greater recognition and support for these programmes from both the public and private sectors. President Clinton recently gave several programmes the first Presidential awards for Excellence in

Microenterprise Development and announced plans to increase funding for the CDFI Fund by $1 billion over the next five years.

One question for the future is how well these new entrepreneurs, still very vulnerable to failure, can survive and grow. Can the alternative development programmes succeed in building the missing ladders of credit and support these entrepreneurs need to achieve stable middle-class incomes? Will there be the child care, the health care, the counselling, the credit and capital, the market opportunities, the temporary income supports, the equipment they may need? How, if they are primarily "locals", can they become "cosmopolitan" and world-class, as Kanter suggests they must become to survive in the competitive conditions of the global economy. If they are "isolates" or "nativists" who fear and resist change, she thinks they have no future. What will it take to change their consciousness? We are a long way from the answers to these questions.

The challenge of the information and technology revolution

Change resulting from the new capacities unleashed by this revolution is already great and more is coming fast. Challenges to women's entrepreneurship will include, among others, understanding and using the full potential of the new tools, getting access to the technology, and responding to changes in business sectors in which women's businesses are concentrated.

In the new digital economy, women, like men, will create wealth by applying knowledge, networked human intelligence and effort to whatever business sector they choose to be in – services, retaining, manufacturing and even agriculture. They will have new access to sources of capital through new computerised investor networks. They will be able to tap the talents of people all over the world. They will be able to do business everywhere. In many important ways they will be able to do more with less and do it more quickly.

This is a new frontier. The players, dynamics, rules and requirements for survival and success are all changing. With the new electronic tools, women's companies (and their competitors) can know more and know it fast. They can be more creative, adding value to their greater knowledge, and creativity thus becomes the new competitive dynamic. Instead of mass production, we will see mass customisation – products and services tailored to individual customer demands, quickly and accurately. In order to keep ahead of the competition, the mass product will be constantly redesigned with added or different features. Product life cycles are already shrinking so fast that the current rule in leading companies is to make their products obsolete as fast as they can.

All companies will be information technology companies, whatever they produce. They will have to be learning constantly, using the computer and the information highway to improve their corporate memory, know what is happening in their world, provide their employees with the information needed to align them with where the company is moving, and help everyone in the organisation to acquire and share knowledge and ideas on a continuing basis. Planning and project development will be able to cover many aspects of a project simultaneously, instead of sequentially, using computer modelling of alternative approaches to a problem. Ideally, owners or employees will no longer have to say: "If we had only known that before we did this, we could have avoided the problem we have now". Companies which have fully integrated information technology into their operations will require new kinds of managers and employees and new relationships between them.

As one commentator on this new frontier has said: "Networked intelligence has been the missing link in organisational learning". Using it, an organisation, like a really centred and integrated person, can become fully conscious of what it is doing, aware of where it has been divided against itself, and able to make better choices to express its vision. There will be an increasing emphasis in business on building "learning organisations", a concept pioneered by Peter Senge in his influential book, *The Fifth Discipline*. As Senge puts it, learning organisations are places "where people continually expand their capacity to create the results they truly desire, where new expansive patterns of thinking are nurtured, where collective aspiration is set free, and where people are continually learning how to learn together."

The new technology and the new thinking places a central emphasis on knowledge workers as the most important asset of business. How many knowledge workers are needed? How many people can be knowledge workers? What happens to those who cannot or do not become knowledge workers? Assuming that most people, given the right education and tools, can become knowledge workers, what revolution in education and training and in access to technology will be required? No one has really dealt with the implications of these questions.

It is already clear that there will be a huge private market for learning products and services. Multimedia training materials which can be accessed through computers with CD-ROM capabilities are already revolutionising business training. Most business training is likely to take place inside businesses using the new technology. Small businesses have mostly provided training in basic employment skills, while large businesses have trained in more specialised ways. The new technology may ease the burden of training on both and enable them to broaden what they offer.

We can expect many scattered and dysfunctional systems such as health care, criminal justice and education, to become integrated and radically changed. In education, experimental schools and universities using the new technology are already providing exciting models. The crucial lesson is that technology alone will not improve schools: it is rather how the technology is used. When used well, it clearly develops independent reasoning, improves writing and research skills and develops confidence and improved self-concept. In such settings, the introduction of training in business skills will be easy and effective. Girls stand to profit from such changes even more than boys because of the way the school experience has hampered their learning and self-confidence in the past.

In the old economy, information flow was physical: cash, checks, written reports, face-to-face meetings. In the new economy, information will become bits stored in computers racing at the speed of light across networks. In the old economy, being big helped to make a company successful. Today, being big is often a liability. Innovation, agility and organisational learning are the key variables for success. Bureaucratic hierarchies are out. The best new enterprises will probably be networks of distributed teams that act as clients and servers for each other. The teams may not even be employees of one enterprise but separate entities which are part of a temporary or permanent alliance to do specific kinds of work. As one observer has put it: "The office will be a system rather than a place."

Information networks will be the highways for the new economy and, we are told, "without a state-of-the-art electronic infrastructure throughout organisations, no country can succeed." For government policy makers, building this infrastructure will have to be a priority.

It is hard to imagine what this coming age of networked intelligence will really mean. Most of it is in its very early stages, although some of it is already operative. One thing seems certain: it will be a world in which there will be no sustainable competitive advantage other than organisational learning. Women's companies – like all companies – will be able to compete only if they can learn faster than the competition. There is both "peril and promise" in that prospect.

Will there be more peril for women than for men? To a great extent that depends on which women's businesses we are talking about. Larger ones can afford the equipment, training and maintenance costs necessary to keep abreast of change. It is the smaller ones that may have trouble getting access to the hardware and software and the support services they need. The budgets of most very small businesses are tight. Without computers and networked intelligence, poor and low-income entrepreneurs may not be able to compete. Access to credit, the diminishing costs of computers and networking, support from non-profit business development programmes and favourable tax policy may be enough to handle the problem, but that remains to be seen.

Several observers have recognised that all businesses based on middleman functions are threatened by how the new technology allows producers and consumers to interact without an intermediary. Wholesaling demonstrates quite vividly how networked intelligence permits direct contacts between buyers and sellers without the need for mediation. In the United States, technology and "connectivity" have allowed alternative distribution channels to develop which have competed successfully with standard wholesalers. The alternative channels include direct manufacturer-to-retailer arrangements with large retail chains, warehouse clubs, discount stores and home improvement centres. There has also been an increase in mail order and catalogue sales, and direct sales from manufacturers to individual customers.

Since women are heavily in middleman businesses – as retailers, wholesalers, distributors, real estate agents, insurance brokers, travel agents, and the like – they will have to shift their businesses to add value of another kind. They will need to move from being in the transaction or information exchange business to providing new combinations of information or services not available on-line.

Practicalities

It is always necessary to come back to practicalities when change is the issue. Therefore, I want to digress somewhat to address an issue applicable to all the conference workshops: policy implementation. It is important to make policy recommendations. But it is just as important to get them adopted and effectively implemented.

Working on issues of women's entrepreneurship both in and out of government, I have learned something about making change happen. Although the following points may seem obvious to such a sophisticated audience, perhaps it will be useful to keep them in mind as we think about the policies we need. I believe they are general enough to have relevance in any democratic country.

◊ *Policy gets adopted when it has an influential champion, when public awareness and support for it appears to be high or when it appears to help a constituency whose favour is sought for political reasons.* Its political appeal is as important as its merit – and sometimes more so. Political appeal is directly proportional to the visibility of the issue and how successfully advocates for the policy have publicised it in the media. Advocates cannot rely only on behind-the-scenes promotion or on the reasonableness of a policy proposal. The perception that there is a powerful women's vote is extremely helpful. Good data showing the importance of women business owners to the economy are invaluable.

◊ *Policy is meaningless without effective implementation.* Mere words, such as the setting of performance goals, not backed by meaningful incentives, enforcement or programme initiatives, are better than nothing because they can be used to pressure for more – but settling for words is almost always a mistake. Where words have had little effect, it is important for advocates to document that and make an issue of it.

◊ *Effective implementation starts with good programme design.* This should involve ongoing collaboration by government with women business owners and their organisations to insure that the design addresses the real problems.

◊ *Good programme design aims at results – outcomes not outputs.* Government frequently sees its function as offering service (output), without focusing on whether the services offered achieve a desired result (outcome).

◊ *To ensure that programmes aim at results, they must be monitored periodically and evaluated annually.* Evaluations must include a good definition of success, appropriate ways to measure success and ways to capture what works best ("best practices"). Common mistakes here are simplistic definitions of success that do not include social as well as economic results, total reliance on quantitative measures and failure to look for the relation between what was done and the varying results. Evaluations should be made available to the public.

- It is now possible to use computers and custom-designed evaluation software to make the collection and analysis of quantitative data uniform and efficient. This software can also improve programme management. It should be augmented by interviews, site visits and narrative reports to obtain qualitative data.

- The importance of evaluation cannot be overemphasized. It is invaluable for programme improvement, for eliminating ineffective programmes and for maintaining government funding.

◊ *Once "best practices" are identified, they should be communicated to programme managers and used to improve programme effectiveness.*

All of this is part of what it takes to make government a useful servant of women's economic empowerment. It is hard work.

Final words

All of the above is a mere beginner's course, a superficial tour of some of the more obvious features of a changing world that women entrepreneurs are already inhabiting. As the new technology spreads, as the institutions of the old economy transform or disintegrate, we will need leaders and policies that address the problems of our national economies in more fundamental ways. A deep moral anxiety is evident in every thoughtful observer of these trends. The threat to basic values is very real, and there is little faith in the ability of governments to protect them.

Yet, we have no choice but to do our best with what we've got – always hoping that there will be a breakthrough before we are out of the game. As we try to hold on to economic growth and create jobs, we are really playing for time, praying we can hold our economies together using the agents of growth and creativity we now have – women entrepreneurs among them – while trying to get ahead of the old game and into the new one.

We can do better. We have talented citizens and many resources in our fortunate nations and we must join together, think together, and build new structures of economic and political co-operation that can keep change from overwhelming us. We can begin building those bonds here, recognising that we must find ways to ease the transition into a transformed world that is coming too fast – and that we must do that with as little damage as possible to the most vulnerable among us.

Chapter 3

THE OFFICE OF WOMEN'S BUSINESS OWNERSHIP IN THE US SMALL BUSINESS ADMINISTRATION: A CASE STUDY

by

Sherrye Henry

Assistant Administrator for Women's Business Ownership

US Small Business Administration, United States

The US Small Business Administration's (SBA) Office of Women's Business Ownership (OWBO) has a 20-year history. ... It was founded in part because of the efforts of committed women advocates and entrepreneurs, and relies even today upon the volunteer initiatives and commitment of hundreds of women business owners and women's organisations around the country. The Office serves as a still-evolving example of how partnerships among government, business and advocacy organisations can work to support and enlarge this vital sector of the economy.

"Women entrepreneurs are a dynamic force in our nation's current economic expansion... [W]omen business owners are the new face of the American and global economy. My Administration will continue to aggressively pursue forward-looking initiatives that will foster the success of these women-owned businesses..." President Clinton, 29 January 1996.

The US Small Business Administration's (SBA) Office of Women's Business Ownership (OWBO) has a 20-year history. From its beginning during the late 1970s, it has served as the only office in the federal government whose sole purpose is to foster women's business ownership. As women's business ownership, in particular, and women's economic self-sufficiency, in general, have become increasingly important national and global issues, the Office has more and more been called upon to serve as the voice of millions of entrepreneurs. In 1994, President Clinton signed legislation making OWBO a permanent office within the SBA and assigning OWBO as the lead federal office on the Interagency Committee on Women's Business Enterprise (a committee of senior representatives from ten federal agencies ensuring that women's economic issues are addressed at the highest policy-making levels). The Office was founded in part because of the efforts of committed women advocates and entrepreneurs, and relies even today upon the volunteer initiatives and commitment of hundreds of women business owners and women's organisations around the country. The Office serves as a still-evolving example of how partnerships among government, business and advocacy organisations can work to support and enlarge this vital sector of the economy.

Overview of OWBO's home: the US Small Business Administration

Who we are and what we do

The US Small Business Administration (SBA) was created in 1953 as an independent agency of the federal government to aid, counsel, assist and protect the interests of small business concerns, to preserve free competitive enterprise, and to maintain and strengthen the overall economy of our nation. Small business is critical to our economy, to building America's future and to ensuring the United States' place in today's global market-place.

The SBA's vision revolves around two principles: customer-driven outreach and quality-focused management. We are determined to reach out to small businesses in an unprecedented way, to listen to their needs, to report these needs back to President Clinton and suggest appropriate initiatives to help small businesses. We also recognise the need to change our management culture, our organisational structure and our business practices to improve the quality of our work. Through these changes, we are creating a more entrepreneurial, customer-driven and efficient SBA. "As champions of the small business community", says Aida Alvarez, Administrator of the SBA, "our agency is working to help build America's future."

Some facts about the SBA, small business in America and American women-owned businesses

The SBA:

◊ Has a portfolio guaranteeing over $29 billion in loans to 200 000 small businesses that otherwise would not have had such access to capital.

◊ Guaranteed over 50 000 loans totalling $10 billion to America's small businesses in fiscal year 1996.

◊ Last year extended management and technical assistance to nearly 850 000 small businesses through its 1 000 Small Business Development Centres and more than 12 000 Service Corps of Retired Executives volunteers.

◊ Provided more than 38 000 loans totalling $987 million to disaster victims for residential property, personal property and business losses in fiscal year 1996.

◊ Has 7 000 private sector lenders as partners providing their capital to small business.

◊ Has increased its venture-capital programme with more private capital in the past two years than in the previous 15 years combined.

◊ Provides loan guarantees and technical assistance to small business exporters through US Export Assistance Centres in 15 cities.

◊ Can respond to written small business questions through the US Business Advisor on the Internet (http://www.business.gov).

US small businesses:

◊ Employ more than 50 per cent of the private workforce.

◊ Generate more than half of the nation's Gross Domestic Product.

◊ Are the principal source of new jobs.

American women-owned businesses:

◊ Are 8 million strong, and make up more than one-third of all businesses.

◊ Contribute $2.3 trillion to the US economy.

◊ Employ one in five American workers.

◊ Are opening at twice the rate of other businesses.

Historically, the SBA has served small business principally through major in-house programme divisions organised to provide financial assistance, training and counselling, assistance with obtaining government contracts and subcontracts, and research and advocacy. SBA guarantees several kinds of loans and revolving lines of credit provided by co-operating banks and non-bank lenders to small business. It guarantees bonds for small contractors, helps to finance Small Business Investment Companies that make equity investments or equity-like loans available to small businesses, provides guaranteed debentures to non-profit certified development companies which enable them to make larger, long-term, job-creating loans to small businesses, and makes disaster loans to afflicted homeowners and small businesses on an emergency basis. SBA relies on "resource partners" which it funds to do much of the training and counselling. These resource partners include Small Business Development Centres that provide technical, assistance and business training; Business Information Centres that provide technological resources; the Service Corps of Retired Executives, more than 12 000 volunteers who provide free counselling for America's entrepreneurs; and Women's Business Centres that provide business skills training and counselling to women entrepreneurs. The agency expects all its programmes in the field to work together in an integrated way.

SBA's policies and regulations are set and overseen by a Central Office in Washington which operates through Regional Offices in the ten Federal regions and through District Offices distributed throughout the population centres in these regions. There are currently ten Regional Offices and 69 District Offices supplemented by two SBA loan-servicing centres and four disaster loan service centres. The regional and district offices are organised into the same major operating divisions that exist in the Central Office and include special staff to aid minority enterprise development. Most of the District Offices have appointed a staff member to be a Women's Business Ownership Representative.

History of SBA's Office of Women's Business Ownership

The SBA's Office of Women's Business Ownership was established in 1979 under an Executive Order issued by President Jimmy Carter.

This Order was a consequence of a research report produced by a federal Interagency Task Force on Women's Business Enterprise. The report, *The Bottom Line: Unequal Enterprise in America*, documented the barriers women were facing in starting and growing their own businesses and called for a government effort to remove those barriers. The President's Order created OWBO and gave it a broad advocacy role. OWBO was charged with the responsibility for promoting federal agency and private sector efforts to assist women-owned businesses into the mainstream of the economy. It was to do this partly by staffing an Interagency Committee on Women's Business Enterprise, comprised of high-level officials from 17 agencies and the White House, who were to promote responsiveness to the Executive Order where they worked. OWBO was also to draw up a strategic plan for the President outlining what other steps ought to be taken.

Shortly after the Executive Order was issued, the Interagency Committee advocated successfully for a pilot mini-loan programme at the SBA which made thousands of loans under $20 000 to women-owned businesses. This responded to the complaints of women that SBA had been unwilling to make or guarantee such small loans, and that the average SBA guaranteed loan of over $100 000 was too much debt for most women to carry. The Interagency Committee also persuaded the SBA's procurement assistance office to do a massive outreach effort to women business owners to register their businesses on SBA's database of businesses interested in doing business with the government. Thousands were registered. OWBO sponsored conferences for women around the country that featured as expert speakers local women accountants, lawyers, bankers and business women, giving these women a visibility and recognition usually denied them at that time. OWBO got extensive media coverage for women's businesses and the new women's business movement, sponsored research on minority women's businesses, committed the US Bureau of the Census to do the first special study of women's businesses, published a guide for women on how to do business with the government, funded a project with the National Association of Bank Women to educate bankers and potential women borrowers about each other, funded a pioneer forerunner of the Women's Business Centres, and did much other outreach to business leaders and business organisations stressing the importance of women's entrepreneurship to the economy.

For the next eight years, OWBO served primarily as an advocate for women entrepreneurs. OWBO's main initiative was to engineer a series of well-attended educational conferences for women entrepreneurs and to produce and distribute at the conferences a number of useful business development guides. The Office catalysed the establishment of several non-profit private sector women's organisations around the country to produce these conferences. Unfortunately, most of these organisations did not continue as independent entities.

In 1988, women's business organisations, frustrated by what they saw as an insufficient response by the Executive Branch to the problems being faced by women business owners, pressed the United States Congress to conduct hearings that would demonstrate the need for stronger government efforts. The hearings indicated that many women-owned businesses were still facing discrimination in access to credit and capital and in getting government contracts and were not getting the kind of business assistance they needed to compete successfully in the market-place. Congress, in response, passed the Women's Business Ownership Act of 1988. Title II of this Act mandated the OWBO to create and run a Women's Business Owner Demonstration Programme. The demonstration was to fund private sector organisations to set up pilot women's business training centres. These were to provide long-range counselling and training to potential and existing women entrepreneurs, including a representative number of socially and economically disadvantaged women.

The legislation gave OWBO, for the first time, programme responsibilities beyond advocacy. OWBO suddenly had the opportunity to show that programmes designed for women entrepreneurs and run by people sensitive to women's needs could have a significant economic impact. With this new mission, OWBO was re-energized, but staff had to learn fast how to administer a programme with awardees around the country. Fortunately, the women's organisations who received the early awards were led by talented and dedicated women determined to make a difference. The Women's Business Centres gave women entrepreneurs high-quality professional services combined with the kind of personalised attention and support that took them seriously, demystified business and helped them overcome self doubt, sabotage by family and friends, and many other difficulties.

After the demonstration programme's first three years, and the favourable preliminary reports to SBA and Congress by the National Academy of Public Administration, a respected independent evaluator, Congress re-authorised the demonstration for another three years, then again for three more years. As the Women's Business Centres proliferated from 12 sites in eight states and the District of Columbia to more than 50 sites in 28 states, the District, legislators and SBA officials became aware of the Centres' contributions at the local level. Satisfied and enthusiastic clients and women's business organisations wrote letters thanking their Congressional representatives for supporting the demonstration. Favourable reports about the programmes and their successful clients in local newspapers and television and radio programmes built community support. In 1995, impressed Congressional leaders doubled demonstration funding from about $2 million a year to $4 million.

As a programme office administering the Women's Business Owner Demonstration Programme, OWBO makes awards averaging $150 000 per year to non-profit organisations that respond to its Request for Proposals on a competitive basis. The non-profit organisations must show how they will provide women with training and counselling on securing credit and capital, financial and general business management, marketing, government contracting and several other optional areas such as production and engineering. OWBO's awards are predicated on a showing of prior experience by the organisation or its staff in providing such assistance and on well-thought-out plans with milestones which show responsiveness to local needs, take advantage of local resources and avoid duplicating existing local programmes. Centres must match the federal dollars at a ratio of 2-1, 1-1, 1-2 federal to private dollars over the three-year period. Centres may charge a fee for their services, but are unlikely to become self-sufficient on this basis because of the large number of low-income women who are provided scholarships and reduced fees. Since 1995, applicants must indicate that they will have installed specified computer equipment that can handle OWBO's new data collection and reporting system and that someone on staff will be responsible for managing that system. The Women's Business Centres programme is an example of one of SBA's most successful public-private sector partnerships, leveraging federal dollars while supporting community-based programmes.

In 1994, President Clinton signed legislation that made the Office a permanent part of the SBA, protecting it from shifts in the political process. The 1994 legislation also authorised the reestablishment of the Interagency Committee on Women's Business Enterprise and restructured and re-established the National Women's Business Council, a public-private partnership advisory council that had been created five years before in the same legislation that authorised the Women's Business Owner Demonstration. The three structures now support each other effectively.

The case for these women's programmes has been greatly bolstered by the statistics developed by the private sector National Foundation for Women Business Owners (NFWBO). In 1994, NFWBO's Compendium of National Statistics on Women-owned Businesses in the United States was published and widely publicised by the Foundation, the National Women's Business Council, the Interagency Committee and OWBO. It showed, to considerable public amazement, that women-owned businesses employed more people than the *Fortune 500*, creating 300 000 new jobs in 1992 while the *Fortune 500* lost 300 000 jobs. This statistic was picked up by national business magazines and by the press and media generally. When figures from the 1992 US Census

Bureau Survey of Women-owned Businesses were released in 1996, the economic importance of women-owned businesses was confirmed. The Census showed the remarkable growth of women-owned businesses in the previous five years a growth that occurred in numbers of businesses, revenues and in all industries. Women are going into business at twice the national rate.

OWBO's advocacy work over the years has been extensive. It has:

- ◊ Helped to form the National Coalition of Women's Business Organisations, a network of over 50 national and local organisations promoting women's business development.
- ◊ Developed a number of informative publications tailored to women or to meeting women's business needs. These include publications on: women's success stories; how to sell products and services to the federal government; how to create a successful mentoring programme; and best practice lessons on running a women's business centre. Also, it produced three directories of non-traditional women-owned businesses for use by federal agencies and prime contractors.
- ◊ Established a Women's Homepage on the Internet – an interactive, multimedia electronic guide which describes what the government can offer to women and provides links to numerous women's organisations that are on the Internet (www.sba.gov/womeninbusiness).
- ◊ Built relationships with women entrepreneurs, women's and minority business organisations, and a variety. of corporations and financial institutions through conferences, meetings, award programmes and individual consultations which ensure that government decision makers understand the day-to-day realities of women's business ownership.
- ◊ Staffed the Interagency Committee's efforts to research what members' agencies can do for women-owned businesses, leveraged resources from those agencies, and helped develop and distribute the Committee's newsletter and videotape.
- ◊ Initiated a Women's Procurement Pilot Programme with 11 federal agencies to expand the pool of women-owned businesses ready to sell to the government in order to support the new 5 per cent national procurement goal for women-owned businesses.
- ◊ Provided support to Women Business Owner Representatives in SBA field offices and helped to train and motivate them to work with nearby women's Business Centres and women business owners.
- ◊ Conceived and publicised the first Women's Prequalification Pilot Loan Programme, now being rolled out in a national pilot, which enables women to prequalify for an SBA loan guaranty before applying to a bank for a loan.
- ◊ Designed and promoted volunteer individual and roundtable mentoring programmes for women now widely sponsored by SBA District Offices across the country.
- ◊ Through speaking engagements and media interviews, publicised women's business successes, their contributions to the economy and its own activities.
- ◊ Successfully promoted more research and development grants to women-owned businesses by the SBA's Small Business Innovation Research programme.
- ◊ Worked successfully with the Avon Corporation to create and produce its annual "Women of Enterprise Awards" programme, which honours women business owners who have succeeded

against great odds and makes them available as speakers, free of charge, to programmes around the country.

◊ Developed a partnership with the IBM corporation to provide all its currently and formerly funded Women's Business Centres with advanced computers and modems to enable them to participate in its new Intranet and Internet activities and to provide access to computers to low-income women.

OWBO initiatives in the "information age"

OWBO has taken two major steps to enter the "information age". The first, "OWBOTrack", enables OWBO to monitor efficiently the performance of the Women's Business Centres, to improve Centres' management information systems, to compare Centres' performance and to evaluate Centres' economic impact. Rather than asking previously funded Centres to revamp existing paper-based systems, OWBO started by introducing the new system in the Centres first funded in 1995.

OWBOTrack's uniform but flexible design and its early installation avoided or eased the problems presented in monitoring and evaluating previous Centres. Centres had used different definitions for basic terms such as what constituted a "start-up" business, collected different kinds and amounts of data, had difficulty doing follow-up on clients or did it on different schedules, resisted filling out and sending paper reporting forms and felt it to be an intrusion on what they saw as their overriding and overwhelming work – doing outreach, providing services to a demanding clientele, fund-raising and managing their programmes. They were so busy serving clients that they were not analysing what was working best or where they were expending significant effort for minimal results. OWBOTrack enables the Centres and OWBO to track performance and make such analyses.

By statute, OWBO is required to submit an annual report to the Committees on Small Business of the Senate and House of Representatives providing answers to five questions regarding the success of WBCs:

◊ the number of individuals receiving assistance;

◊ the number of start-up businesses formed;

◊ the gross receipts of assisted businesses;

◊ the profits of assisted businesses;

◊ the employment increases or decreases of assisted businesses.

Prior to OWBOTrack, OWBO relied upon less detailed data, although it was clear that many thousands of women had been served, and that many start-ups and jobs had been created at reasonable cost. Estimates were that one business and five jobs had been created for every $10 000 in federal dollars expended.

The development of OWBOTrack took time, effort and money. After about 18 months of work with an evaluation contractor and software developer, feedback from the sites, software programming, removing "bugs", and training Centre staff, OWBOTrack is now up and running. It will be continually refined as experience develops, but it is already providing quantities of useful information never available before. There is great interest in OWBOTrack from other programmes in government and the private sector, and OWBO is exploring how best to extend its usage.

OWBOTrack has many exciting capabilities. One is that it can track which clients choose which services and how many hours they receive from any service. Potentially, given good follow-up data, OWBO will be able to analyse what combinations of services, with what intensity, lead to what successful outcomes for what kinds of clients. Both Centre and OWBO staff can easily access and manipulate this and other data for both reports and analysis. The Centres find that using OWBOTrack, they now understand their own operations much better and that this new access to their own information makes many tasks easier.

OWBOTrack is already providing a detailed profile of the clients, their businesses, their problems and their goals, which was not available previously. If already discernible patterns continue, the Centres appear to be serving mostly beginning or aspiring women entrepreneurs, predominantly between the ages of 35 and 55, but with many between 25 and 35, the majority with high school or college educations, about half never or no longer married, and usually having no more than two dependants under the age of 18. They are about equally self-employed or working for others and about 10 per cent are unemployed. They have low or moderate incomes and, if in business, have mostly been in business for less than a year. About 40 per cent of their businesses are facing either uncontrolled growth, instability, shrinkage or problems the owners cannot identify. Their businesses are about 60 per cent in services and about 27 per cent in retail, with other industry categories far behind. Wholesale trade at almost 9 per cent, manufacturing at about 7 per cent and construction at just over 5 per cent, are predominant in the remaining distribution, although all standard industrial categories are represented except mining.

OWBO's second foray into the electronic age is new, exciting and, as yet, not fully developed: it has two parts. Already operating on a test basis is the first part, the Women's Business Intranet (WBI), also known as OWBOZone. It is a private Intranet, entered only with a confidential password, that electronically links all of the Women's Business Centres funded by OWBO with each other and with OWBO. When fully developed, this will facilitate the exchange of information that has proven so helpful at the annual in-person briefings for the Centres and will enable informal exchanges to be available continuously. In the OWBOZone there will be chat rooms, interactive conferences, topic forums, an e-mail centre with broadcast e-mail capabilities, and sharing of curricula and best practices. This should improve the productivity and effectiveness of all the Centres. OWBO will be able to post notices and reminders, respond to queries and receive reports through this linkage, reducing its need to send faxes and mailings and make telephone calls.

The second part is the anticipated Online Women's Business Center. This will be a virtual training and counselling centre existing in cyberspace. When fully developed, it will offer information, interactive training in entrepreneurship, access to counsellors, links to existing "live" centres and other government and private sector resources, electronic bulletin boards, chat rooms, conferences and forums to anyone on the Internet wherever they live, every day and all day.

In 1996, after a competitive process, OWBO funded the North Texas Women's Business Council (one of the 53 centres) to develop this Online Center. Each of the existing Centres have received one year's funding to help build the Online Center, in particular by using their own most successful programmes as a basis for the Online curriculum.

Designing this Online Women's Business Centre is a complex and novel undertaking which will start first with simple offerings sometime in the summer of 1997. As response to the website grows and comments are received from users, changes and additions to content will be made.

How to evaluate the impact of the Online Centre is another novel challenge. It is OWBO's intention to solve problems as they arise and to be at the leading edge of the new technology to spread the knowledge its Centres have acquired and to empower women entrepreneurs to learn from each other and from other resources.

Policy lessons

◊ In a time of declining federal spending and increased focus on public-private partnerships, leveraging federal resources through women's business coalitions, organisations and extraordinary individuals increases the effectiveness as well as the stability and continuity of government efforts.

◊ Community-based training organisations, supported by federal resources and best practices, but responsive to local needs, population and demographics, can make a tremendous difference in the likelihood for an aspiring women entrepreneur's success.

◊ Successful government programmes for women entrepreneurs must take advantage of changing technological resources to ensure that the programmes reflect the most current available technology and that the women served are versed in the technology enhancing and directing the growth of small business in the "information age".

Chapter 4

THE ROLE OF NETWORKS

by
Amy Millman
Executive Director, National Women's Business Council
United States

> *The women's community must organise a network to create its own agenda of economic security for women. The leadership of this network must work in partnership with the public and private sectors to help women start and expand businesses with training and technical assistance, access to credit and capital, marketing opportunities including government contracts, and research to develop a profile of these businesses.*

In the increasingly competitive global economy of the 21st century, we must leverage all of our entrepreneurial resources if we are to achieve our greatest economic potential. Women represent a majority of our population, and we are now entering and succeeding in the business world in unprecedented numbers. As of 1996, there were nearly 8 million women-owned businesses in the United States, employing more than 18 million people and generating more than $2.3 trillion in sales. Women-owned firms account for one-third of all firms in the country, provide employment for one out of every four US workers, and generate 16 per cent of the nation's business sales. By the end of this century, the projected growth in the number and maturity of women-owned enterprise will continue to accelerate rapidly.

The growth of women-owned enterprises in the last decade has been a driving force in the revitalisation of the American economy. Their contributions have been duly recognised and the Executive Branch of the United States federal government, working closely with its Congress, is aggressively pursuing forward-looking initiatives that will actively assist more women in starting and growing their businesses. In order to build a strong infrastructure that supports entrepreneurial development, the government must involve all the nation's resources, both public and private, in this effort. Consequently, women, who will own more than 40 per cent of the businesses and comprise more than 60 per cent of the workforce by the year 2000, are being asked to play an increasingly important role in fashioning a new economic blueprint.

Building an infrastructure to foster women's business development

As evidence of the increasingly powerful role business and professional women and their networks play in the overall health and welfare of our economy, Congress and President Clinton established, in 1995, a public-private partnership intended to foster women's economic development. President Clinton appointed five prominent business owners and five leaders of national women's business organisations to the National Women's Business Council.

To serve on the public sector panel, the Interagency Committee on Women's Business Enterprise (Committee), he appointed senior federal government policy makers. Together, they link women business owners, women's business organisations and senior government policy makers in a bold new initiative to promote the success of women's business enterprise nation-wide.

The partnership

The National Women's Business Council (hereinafter, the Council), a bipartisan advisory panel of the federal government, was authorised by Congress in 1994 to serve as an independent source of advice and counsel to the President, Congress and the Committee, on economic issues of importance to women business owners and on the effectiveness of programmes and policies designed to support women's business development. With the Council, President Clinton sought to establish a direct network for communication and advice from the experts – the business owners and advocates themselves. President Clinton appointed one of the most successful self-made women business owners in the country to chair the Council – Lillian Vernon, Chair and CEO of the Lillian Vernon Corporation. To spread the message that government was offering women entrepreneurs a seat at the policy table, four other prominent business women and five leaders of national women's business organisations were chosen to comprise the Council. Together, the Council's network represents the voices of several million women entrepreneurs to the economic policy makers of the United States.

Top government officials were appointed to the Committee to ensure that women's economic development was a top priority in the policies and initiatives of every federal agency. Committee members represented the National Economic Council, the Executive Office of the President, the Small Business Administration, the Departments of Commerce, Defense, Health and Human Services, Labor, Transportation, Treasury, the General Services Administration, and the Federal Reserve, the leading financial regulatory agency.

The innovative partnership of the Committee and the Council has resulted in a new energy, new vision and new understanding of the importance of including women's voices in the policy arena shaping the economy of the 21st century. Prior to this, women's economic development had not been integrated formally into the federal government's priority list. But, with the creation of this partnership, not only did women business advocates have a channel through which to voice their concerns and recommendations to the policy makers of the Committee, but this Interagency Committee became both a catalyst for change and a vehicle to market the federal government's portfolio of policies and programmes that promote greater opportunities for women. The infrastructure was now in place for the 700 women's business organisations throughout the nation, representing 8 million women business owners, to access information, evaluate existing programmes, and make their visions and priorities a part of the economic decision-making process.

Expanding the network

At its inaugural meeting in January 1995, the Council held a plenary session at which leaders of the top 30 women's business organisations were invited to share their priorities and to recommend policy initiatives that could be communicated through the new public-private partnership and implemented by the federal government. The members of each organisation agreed to use their resources and expertise to assist the Council in accomplishing four objectives critical to women business owners: expand and ensure access to financial resources; improve networking, technical assistance and training and the infrastructure that supports them; expand competition by opening public and private market opportunities; and create a comprehensive profile of the women-owned business sector through research and data collection.

At the same time, Interagency Committee members pledged to use the federal government's infrastructure and resources to address women's economic concerns for expanded opportunities and economic security. The Council and the Committee adopted a two-pronged approach to achieving their objectives. Building on the interest expressed at the first meeting, the Council forged a vast nation-wide network to give women a strong voice in Washington while enabling the Council to effectively communicate the challenges women faced in starting and growing their businesses. The Committee expanded its ranks to include officials from every federal agency, and established a system to communicate and co-ordinate their shared agenda and models of programmes that were working to promote women's economic development. Together the Council and the Committee created and participated in meetings of four issue task forces. The task forces executed a top-to-bottom review of existing federal policies and programmes for women entrepreneurs administered in each federal agency. The results of their work were issued in a January 1996 report to President Clinton and Congress. The report describes partnerships that have already been forged and points the way to future collaborations. In her official presentation, Chairwoman Lillian Vernon pronounced that this "public-private partnership is the foundation for the greater involvement that we envision for women entrepreneurs in setting the economic agenda for the nation."

Through the network established by this unique partnership, 8 million women business owners no longer operated in isolation. The network opened a direct line of communication between the federal government and the women's business sector. It also provided women business owners with the most valuable asset of all – access to information, new resources and a whole world of new opportunities with which to grow their enterprises. The success of this new advocacy is best illustrated through the stories of the women who have benefited.

A letter of introduction

Susan Corrales-Diaz was having great difficulty building her business in the United States. As President and CEO of Systems Integrated, an innovative high-technology company that manufactures automated control systems for the Defense Department and electric utility companies, she was competing for contracts in a depressed market. Since the late 1980s, the economy had been in recession, the federal government was under intense pressure to cut costs, close bases and eliminate the bulk of its contracts – making the competition for the remaining lucrative contracts even more fierce. Ms. Diaz found more often than not that these contracts were awarded to the largest companies. She could not compete. Searching for new markets, she approached the federal government for assistance in gaining access to the international market-place. Without a "letter of introduction", small businesses had little chance of being taken seriously by other governments. To gain an equal footing she needed not only her proven record of efficiency, flexibility and reliability, but also an official indication that the US government supported her efforts. Through her contacts with the Council, she was referred as a potential candidate to join the Secretary of Commerce on a series of trade missions that he would personally lead. Not only was she given a "letter of introduction" and access to a new world of business opportunities – rarely available to small enterprises – but her experiences were translated into additional assistance in the areas of technical assistance and financing for these new overseas ventures. Ms. Diaz now serves on the President's Export Council, a prestigious group of business leaders who have a major impact on developing the infrastructure that actively supports and advocates for expanded opportunities in the global market-place for ALL businesses.

From flower pots to fighter jets

In Seattle, Washington, the global market-place was also having an impact on the business interests of Grace Tsujikawa Boyd, President and CEO of Pyro Media, Inc. In ten years, Ms. Boyd had built a solid reputation for producing the high-end, quality ceramic planters found in most of the elegant office buildings and hotels world-wide. However, low-cost foreign imports were beginning to price her out of the market. Through the Council's network, she discovered and took advantage of a government-sponsored mentor-protégé programme, which provides incentives to large businesses who provide technical assistance and training to small businesses. This partnership combines the technological facility of the large contractor with the specific expertise of the small business to produce goods and services for the government more cost-effectively than they could independently. Through the Air Force's mentor-protégé programme, teaming them with Boeing, a large aerospace company who was searching for new materials to produce parts for its F-22 jets, Pyro Media's product line expanded from flower pots to fighter jets. Not only did this programme provide Pyro Media with a significant foothold in an important new market, Boeing was able to fulfil the terms of its contract at a significant cost savings to the Air Force.

In both instances, the network provided these women with the necessary information which enabled them to take advantage of new competitive opportunities to grow their businesses. The Council and Committee's new infrastructure has been remarkably effective – both in recognising the concerns and needs of women business owners and providing them with the means to change dreams into reality.

Setting an economic agenda for the future

With the network in place and significant action occurring throughout the federal government addressing women's economic development, the Council began to chart new territory: the creation of a new infrastructure that would support the growth of women's businesses well into the 21st century.

Involving the new expanded network was the first priority. In May 1996, at a prestigious academic business institution, the Council and the Committee hosted SUMMIT '96 – a national women's entrepreneurial summit. More than 500 women business owners, leaders of women's business organisations and officials from corporate America, academia and government gathered to consider and debate an agenda that will foster women's business ownership and create a new economic vision for the 21st century.

The Summit '96 Action Agenda

Create innovative approaches to financing businesses. Increase the flow of capital and create new financial products and services to address the needs of the new economy, including those that make better use of secondary markets and pension funds as sources of capital.

Encourage investments in entrepreneurial training. Create programmes that expand and strengthen women's entrepreneurial skills.

Promote opportunities/alliances in the global market-place. Facilitate their participation in commerce and embrace initiatives that have improved the economic conditions of women and families in other countries.

Increase competitive market opportunities. Increase competitive access by women-owned businesses to the public and private procurement markets, establishing as a goal to make their participation commensurate with their numbers by the year 2000.

Build technological capacity. Develop and introduce mechanisms to build the capacity of women entrepreneurs to access and utilise technology.

Promote best practices inside and outside the corporation. Women have fashioned a model that is flexible, streamlined, "friendly" and responsive to workers, communities, the environment and the new economy. It should be studied and applied wherever appropriate.

Construct a foundation of economic literacy for young women. Make economic literacy and economic empowerment for women and girls integral to course material and teaching in our education system.

Redefinition of work. The statutory, regulatory and corporate definition of work must include self-employment as a viable career and employment strategy.

Summit leaders endorsed eight policy statements which, along with the accompanying business initiatives and research topics, formed the basis for a women-led economic agenda. This agenda marked a milestone, not just for women business owners but for the entire economy. It reflects an economic vision for women and for all Americans by addressing basic issues such as building the capacity of women business owners by investing in entrepreneurial training, promoting best practices inside and outside the corporation, and constructing a foundation of economic literacy for young women. This agenda addresses the challenges of the future. It is the benchmark by which progress will be measured and by which history may well date the "new American economy".

Implementation of the Summit agenda

"The collaboration begun at the summit needs to continue. Now our challenge is to move swiftly to implementation", said Lillian Vernon, Chair of the National Women's Business Council. Using the entire network infrastructure – through the federal government, state and local agencies and business advocates in collaboration with local business, social, philanthropic and community organisations – the Council set out to market this agenda and it is firmly woven into the very fabric of the economy. Over the next few years our job is to bring these initiatives to the attention of the government and business leaders who will assist us in taking the next steps towards their implementation.

Implementation began almost immediately for many of the SUMMIT '96 initiatives.

◊ *Creating Innovative Approaches to Financing Businesses* was SUMMIT '96 Agenda Item No. 1. The participants who helped fashion this initiative urged lenders and investors to facilitate the flow of capital to small businesses to create new products and services to fill the current gaps in the continuum of funding sources. Among the many innovations identified was the creation of an on-line nation-wide network of "angels" interested in investing in the growth of small enterprises.

◊ In October 1996, President Clinton launched a new Internet-based listing service, ACE-Net, which provides information to investors on small, dynamic, growing businesses. ACE-Net is a government-sponsored matchmaking service produced by the US Small Business Administration. It was created in direct response to requests from growing US companies: to find potential investors who can provide seed capital of $250 000 to $5 000 000.

◊ Innovative companies need investment capital to grow. In the United States, entrepreneurs have traditionally been able to obtain financing from family, friends or banks for loans up to $250 000. Above the $5 million level, venture capital has been available. Between the two figures, however, has largely been the province of "angels" – wealthy individual investors with business savvy. While there is a large demand for that level of financing, until now there was no national vehicle designed to bring angel and entrepreneur together. With a password, accredited angels will search on-line through a database to locate a company that meets their specific criteria.

◊ This system presents an opportunity for the Council to encourage investments in women-owned firms. ACE-Net will be an invaluable tool as we begin to prepare women business owners for growth and to groom potential investors by introducing them to the women's business sector.

◊ Preparing all America's business men and women, large and small, to successfully grow their businesses has become an economic policy priority for the United States Government. Building the capacity of businesses to compete for new opportunities both in the domestic and global markets requires greater investments in training, technical assistance and advocacy than our government had previously thought necessary. We understand that, as we move into the 21st century, our intervention is critical to ensure a strong US economy, in which all contribute to their greatest potential and all have a strong voice in shaping it.

Conclusion

As the 21st century approaches, all nations must begin to fully utilise their most important resources – human resources – to succeed in the growing, competitive, high-tech global economy. Small business is essential for achieving a strong, prosperous national economy. Since women own and are starting and expanding substantial numbers of small businesses – 8 million, comprising 36 per cent of all businesses in the United States – addressing the special needs of women's business ownership is essential to the overall economic contribution of small business.

Creating an infrastructure and developing and implementing economic initiatives that foster women's business ownership requires participation from three distinct sectors working in partnership: the public sector, including all branches of the government; the private sector, including profit-making corporations, non-profit organisations and educational institutions; and the women's community, including advocates and entrepreneurs organised to set an agenda for their economic development.

The women's community must organise a network to create its own agenda of economic security for women. The leadership of this network must work in partnership with the public and private sectors to help women start and expand businesses with training and technical assistance, access to credit and capital, marketing opportunities including government contracts, and research to develop a profile of these businesses.

The leadership governing a nation must support the underlying premise of women's economic security to ensure the development of an infrastructure and policy initiatives that foster women's business ownership. The government must accept and acknowledge the recommendations of the women's community by creating model programmes that work for the special needs of women.

The private sector must also acknowledge the contribution and foster the success of women-owned businesses. Including women in leadership positions by giving them a seat at the decision-making table within their own corporations, organisations and educational institutions enhances the expertise of the debate and the inclusiveness of the policy directives.

Promoting women's business ownership is a responsibility shared by the women's community, the private sector and the nation's governing bodies. In the United States, the National Women's Business Council on behalf of a coalition of women's advocacy organisations, the Interagency Committee on Women's Business Enterprise on behalf of the federal government, the US Congress and representatives from the private sector are working in partnership to prepare our nation for a future that will depend on the contributions of women for economic security and prosperity.

The National Women's Business Council submits the following three recommendations to foster women's business ownership and strengthen women's economic security:

Recommendation 1: Appoint an advisory council of leading women's entrepreneurial experts and advocates and a committee of highest ranking female government officials to promote economic initiatives for women entrepreneurs and to create a co-operative effort in building a private and public sector network to foster women's economic self-sufficiency.

Recommendation 2: Obtain official action from the sovereign government to recognise the contributions of women-owned businesses and to promote greater access to opportunities for these businesses.

Recommendation 3: Convene an assembly of business leaders, women in business, officials from academia and government to develop an agenda of key initiatives that foster women's business ownership. Create a plan for implementing and measuring the results of these initiatives.

Chapter 5

THE NATIONAL ASSOCIATION OF WOMEN BUSINESS OWNERS

by
Suzanne Taylor
The National Association of Women Business Owners, United States

NAWBO's vision is to propel women entrepreneurs into economic, social and political spheres of power world-wide. Our mission is to strengthen the wealth-creating capacity of our members and promote economic development; to create innovative and effective change in the business culture; to build strategic alliances, coalitions and affiliations; and to transform public policy and influence opinion makers.

The National Association of Women Business Owners represents the nearly 8 million women business owners in the United States. We generate $2.3 trillion in sales per year and employ 25 per cent of the domestic workforce employed in the private sector. NAWBO's vision is to propel women entrepreneurs into economic, social and political spheres of power world-wide. Our mission is to strengthen the wealth-creating capacity of our members and promote economic development; to create innovative and effective change in the business culture; to build strategic alliances, coalitions and affiliations; and to transform public policy and influence opinion makers.

The National Association of Women Business Owners was formed in 1975 by a group of women business owners in the Washington, DC area. These visionary women realised they needed to network with one another to share insights and experiences, and could find no formal organisation through which to do so. Therefore, they met informally with one another for several months. These meetings eventually formalised, and NAWBO received its articles of incorporation and federal tax status in July 1975.

NAWBO became much more than a networking group, we became a powerful grassroots advocacy organisation. Within a couple of years NAWBO formed chapters across the country and was thriving as a business women's association. Today we have over 60 chapters across the United States and over ten chapters in formation.

NAWBO's visionary early leaders understood the need for a powerful advocacy organisation representing women business owners. In 1977 NAWBO began its successful path of active participation in influencing policy changes at the federal level by testifying before the Senate Small Business Committee. NAWBO continued to advocate on behalf of women business owners and small businesses, and testified before Congress again in 1980 and 1983.

As the leaders of NAWBO learned how effective the organisation was in regard to effecting change in public policy for the benefit of women business owners, we formalised our advocacy efforts and began to train our members in the importance of their voice as women business owners. In 1983 NAWBO held its first Public Affairs Days in Washington, DC, where we began a sound platform of public awareness and training in lobbying for our members. Our Public Policy Days will be held this year in Washington, DC, from 25-29 October 1997. With the ten-year anniversary of the passage of HR 5050 – the Women's Business Ownership Act – on the horizon, we find it to be the perfect time to honour the accomplishments of this legislation. During Public Policy Days we will educate our members on key issues impacting our businesses, continue training on how to advocate small-business issues before their Congressional delegations, and we will bring Congressional leaders to our events to brief us on the 105th Congress.

One of the largest contributors to NAWBO's success in affecting economic policy is through the statistics provided by our sister organisation, the National Foundation for Women Business Owners. Many of the same visionary leaders who helped to create NAWBO also created our Foundation. The National Foundation for Women Business Owners performs research both domestically within the United States, and also internationally. It provides government officials, business leaders and members of the academic community with research and statistics revealing the dynamics and strength of the women business owner segment of the economy, allowing us to collectively effect change on economic policy.

NAWBO was founded by visionaries and throughout our history we have made visionary decisions. In 1984, long before the other American small-business groups were focusing on international business and the global economy, NAWBO joined as the US representative to *Les Femmes Chefs d'Entreprises Mondiales*. This affiliation resulted in NAWBO's participation in trade missions to London, Frankfort, Madrid, Tokyo and Mexico. NAWBO was one of the few small-business organisations to advocate for the passage of NAFTA. Our Foundation is now gathering data and statistics internationally, we will begin to see the same impact for women business owners around the world.

In 1986 NAWBO mounted a grand showing of our forces at the White House Conference on Small Business – 26 of our 27 issues were adopted. NAWBO strategically motivated its membership and led its forces for the 1995 White House Conference on Small Business. NAWBO had more than 300 delegates to the Conference – more delegates than any other association. In preparation for the White House Conference, NAWBO held two "think-tanks" with NAWBO members, policy experts and pioneering practitioners, in such fields as technology, human capital, labour law, community development, procurement, access to capital and regulatory reform. From these think-tanks NAWBO garnered the best recommendations and amalgamated them with the ideas from leading futurists to create the organisation's *2020 Vision: Entrepreneurial Policies for the 21st Century.* NAWBO used these policy recommendations to spur debate and help shape the White House Conference on Small Business. NAWBO urged its delegates present at the Conference to be bold and innovative; it was a leader in the policy development and subsequent implementation efforts of both Conferences. Today, NAWBO is actively involved in bringing to fruition the recommendations of the 1995 White House Conference on Small Business.

NAWBO's greatest achievement in effective advocacy work was its effort to drive the 1988 Congressional hearings which resulted in the passage of HR 5050 – the Women's Business Ownership Act. HR 5050 was a monumental piece of legislation that NAWBO's late Past National President, Gillian Rudd, helped to spearhead on NAWBO's behalf. NAWBO recognised the need for women entrepreneurs to be recognised as a driving force in the US economy and demanded the attention of the United States government. HR 5050 created the Women's Demonstration Sites, supervised by the Office of Women's Business Ownership at the US Small Business Administration. These demonstration sites, now referred to as Women's Business Centres, assist thousands upon thousands of successful and aspiring women entrepreneurs to realise their dreams. NAWBO was instrumental in obtaining the Congressional hearings, and helped to orchestrate their wonderful success – as evident today in the activities of the Office of Women's Business Ownership as well as the National Women's Business Council, which was another outgrowth of HR 5050.

During the first session of the 105th Congress of the United States in 1997, NAWBO testified on the issues of women business ownership, procurement and the 5 per cent goals programme in the federal government for procurement contracts to women-owned businesses. Currently only 1.8 per cent of all federal contracts are awarded to women business owners. NAWBO finds this level unacceptable, and is actively advocating that the federal government consider more women in procurement opportunities.

In April 1997 NAWBO will testify before the Senate Committee on Small Business on the issue of the historic HR 5050 legislation. It has been nearly nine years since this legislation was enacted, and NAWBO will present testimony to help determine what has been accomplished thus far, those aspects of the legislation that are no longer working, and to help determine where we go from here. NAWBO has asked the Committee to revisit HR 5050 and we look forward to working with the Committee on our mission and vision for women business owners.

NAWBO provides a strong and continuing voice and vision for women business owners within the economic, social and political communities. NAWBO exercises leadership to create a business climate which realises the integration of values, profits, technology and people. And, finally, NAWBO seeks to assist with the election of the first woman business owner to become president of the United States, and to assist with providing her with a Congress full of women-owned businesses.

Chapter 6

INVESTING AND DOING BUSINESS OVERSEAS: AN AUSTRALIAN PERSPECTIVE

by
Angeline Low
Managing Director
Ken Resources Pty Ltd, Sydney, Australia

Interviews with ten successful women entrepreneurs were conducted for this report. They provided examples of the kinds of strategies they used to develop export markets and to do business and/or invest overseas. Some entered into strategic alliances (such as joint ventures), some into informal partnerships, some used distributors in the overseas country, some relied heavily on personal contacts and others on showcasing opportunities such as trade missions and exhibitions.

Relatively few women finance their export activities through financial institutions. Although there is no evidence of discrimination by banks in their lending practices, women have a perception that they are disadvantaged in their dealings with them.

Women believe they have particular skills that advantage them in dealing with people in business and government in other countries. These include communication, relationship-building and cross-cultural sensitivity. However, women do face barriers. In particular, they have to balance their entrepreneurial activities with their family responsibilities and they have to deal with cultural differences in attitudes.

Women in small and medium-sized businesses in Australia are globalising, but it is not known to what extent. Estimates suggest that between 10 and 30 per cent of women-owned businesses are exporting. Women make up one-third of people working in their own businesses and they are over-represented in three industry divisions: Community services; Recreational, personal and other services; and Retail trade. To what extent women in these service industries are vigorous exporters or active international traders is not known. Prospects are good for export growth in small and medium-sized service businesses, although there is some evidence that women in small businesses prefer to stay small.

Use of government sources of business assistance is generally low. There may be a number of reasons for this, including the unmet special needs of women such as child care and a strong preference for an all-female environment. Migrant women may feel more comfortable relying on personal and business networks in their home country.

The Women in Export programme operated by Austrade (the Australian Government's international trade advisory service) aims to ensure that women in business are aware of and have access to the full range of Austrade's services.

Interviews with ten successful women entrepreneurs were conducted for this report. They provided examples of the kinds of strategies they used to develop export markets and to do business and/or invest overseas. Some entered into strategic alliances (such as joint ventures), some into informal partnerships, some used distributors in the overseas country, some relied heavily on personal contacts and others on showcasing opportunities such as trade missions and exhibitions.

Relatively few women finance their export activities through financial institutions. Although there is no evidence of discrimination by banks in their lending practices, women have a perception that they are disadvantaged in their dealings with them. One approach may be to increase women's access to finance through new business partners by extending business networks and mentoring programmes. This approach recognises that men typically draw their partners from active networks and business contacts.

Women believe they have particular skills that advantage them in dealing with people in business and government in other countries. These include communication, relationship-building and cross-cultural sensitivity. However, women do face barriers. In particular, they have to balance their entrepreneurial activities with their family responsibilities and they have to deal with cultural differences in attitudes. Doing business in Japan and Korea presents special problems for women.

Women comment on their lack of political influence in business policy making and on the insufficient recognition of their role in the Australian economy. The Australian Council of Businesswomen Ltd (ACOB) has been formed to develop and comment on policy initiatives, to lobby for effective measures that will enhance the ability of women to compete and to promote the achievement of women in business.

A number of recommendations are put forward and these appear below.

Introduction

Many Australian business women are involved in globalisation and international trade through their own small and medium-sized enterprises. In this paper the following activities are included in what has been termed investing and doing business overseas:

◊ exporting goods and services overseas;

◊ other international trade, including outsourcing of production and sourcing of raw materials overseas;

◊ formation of strategic alliances for product development, production, sourcing and marketing;

◊ import replacement and competition with overseas firms in Australia's domestic market;

◊ investing overseas.

To supplement the limited research on this topic, interviews were conducted with ten Australian women entrepreneurs who are globalising. Information from this interview survey is incorporated into the material brought forward from relevant Australian research to address some of the themes of this paper.

While some of the issues facing business women who are globalising are not necessarily gender-related, others are.

Wherever appropriate, recommendations for changes that will encourage women to venture overseas are developed and highlighted.

Extent to which women owners of SMEs are globalising

Limitations on data

There is a lack of statistical information in Australia on women-owned SMEs investing and doing business overseas. In general, data that relate to SMEs doing business overseas are not gender-specific. Similarly, gender-specific and gender-comparative studies of the owners of SMEs in Australia do not include information about the international component of the SME. This may indicate the relatively low expectation of researchers that internationalisation is an emerging feature of, and an issue for, SMEs.

While there are many women who are pioneering international trade, an internal report of the Australian Council of Businesswomen Ltd (ACOB) found there were no specific statistics available on the exporting activities of Australian women entrepreneurs (ACOB, 1997).

Australian SMEs

It is not known whether women-owners of SMEs are more or less aware of globalisation than their male counterparts. We do know, however, that about 4 per cent of Australian businesses have an export component (Industry Commission/Department of Industry, Science and Tourism, 1997), compared with 30 to 40 per cent of small businesses in Asian countries (McMahon, 1997). It appears that Australian businesses, in general, are less involved in exporting than their Asian counterparts.

Manufacturing SMEs are more likely to be involved in exporting – an average of 14 per cent across all manufacturing sectors. By contrast, about 3 per cent of service SMEs are engaged in export (Industry Commission, Department of Industry, Science and Tourism, 1997)

Findings from interviews with women entrepreneurs

To be included in the survey the women interviewed for this report all had to meet the criteria of having some degree of involvement with international trade. Each of the respondents proved to be very much aware of globalisation in some form or other as a business imperative:

- ◊ six were exporters: the proportion of their total revenue from exports ranged between 1 and 90 per cent;
- ◊ one had exported in the past, but was no longer exporting;
- ◊ one was actively pursuing opportunities to extend her business into South-East Asia;
- ◊ one did not export her products, but was very concerned about competition in her local market from overseas firms and was actively seeking to replace those imports;
- ◊ one was an importer who traded overseas.

Target countries for export and investment were located in:

- ◊ the Asian region: Malaysia, Singapore, Thailand, Japan, China, Indonesia and Vietnam;
- ◊ Western Europe and the United Kingdom;
- ◊ North America;
- ◊ South Africa.

The few women who were investing overseas were doing so primarily to support their export business. Overseas investment took the form of distribution and licensing arrangements, property acquisition for office accommodation or production facilities, and joint venture partnerships, both formal and informal. Investing overseas was seen as an integral part of the export business rather than a financial investment.

For migrant women from South-East Asia there was some pressure to invest in family businesses back home. One migrant woman said she never invested overseas, not even with family businesses, because she was too remote to protect her investment. However, she continues to form strategic alliances with foreign manufacturers to make her products according to technical specifications required in Australia.

While the extent of globalisation of the businesses varied greatly, all commented on the strategic importance of interacting with the global economy.

Findings from other relevant research

A recent telephone survey of 44 Australian women entrepreneurs found a high proportion involved in exporting (Ernst and Young, 1996):

- ◊ 93 per cent owned their own businesses;
- ◊ two-thirds of the businesses employed 20 persons or fewer (*i.e.* are small businesses);
- ◊ one-third were involved in exporting;
- ◊ two-thirds of those who did export saw that part of their business becoming more significant.

Factors in global awareness

Some women seem unaware of global opportunities or of the potential for competition from overseas firms in their own domestic market. Those women who are aware may wish to keep their businesses small so that they can retain their autonomy and simplicity. A recent report by the Tasmanian Women's Consultative Council (1996) concluded that Australian research shows that women are more likely than men to adopt a deliberate strategy to remain small rather than pursue a high-growth strategy (Tasmanian Women's Council 1996 with reference to National Board of Employment, Education and Training, 1994).

Distribution of women entrepreneurs by industry sector

Australian statistics

Statistics are available on the numbers of women and men working in their own businesses in Australia broken down by industry. Table 1 shows there are approximately 200 000 women working in their own businesses and almost half of them are contained in just two industry divisions: Retail trade; and Recreational, personal and other services. The data do not indicate the extent to which these women are involved in globalisation.

Table 1. Women working in their own businesses by industry division

Industry division	Women	%
Mining	600	0.2
Manufacturing	17 900	8.7
Construction	19 000	10.1
Wholesale trade	10 000	4.8
Retail trade	51 600	28.1
Transport and storage	7 600	3.7
Finance, property and business services	31 900	12.8
Community services	28 000	11.6
Recreational, personal and other services	37 000	20.1
TOTAL	**203 600**	**100.0**

Source: Australian Bureau of Statistics, 1993, reported in The Flinders University, 1995.

The same statistics (Table 2) show that women make up one-third of people working in their own businesses throughout Australia and that they are over-represented in three industry divisions:

◊ Community services (55 per cent);

◊ Recreational, personal and other services (48 per cent);

◊ Retail trade (41 per cent).

This concentration of women in traditional "female" service industries is not unexpected. The extent to which women in these industries are investing or doing business overseas is not known.

Table 2. Persons working in their own business: percentage of women in each industry division (1993-94)

Industry division	%
Mining	12.2
Manufacturing	34.2
Construction	13.7
Wholesale trade	31.7
Retail trade	41.4
Transport and storage	15.5
Finance, property and business services	29.1
Community services	55.1
Recreational, personal and other services	48.1
TOTAL	**31.8**

Source: Australian Bureau of Statistics, 1993, reported in The Flinders University, 1995

Findings from relevant research

Table 1 shows that less than 10 per cent of women working in their own businesses are involved in manufacturing. By contrast, of the ten women interviewed for this report, six were involved in export and half of those were engaged in some form of manufacturing enterprise. This result is limited by the small numbers of women interviewed, but is in line with the over-representation of manufacturing in the Australian export arena and the under-representation of the services.

Key findings of a 1994 report prepared for the Australian Government on the internationalisation of Australian SMEs in the service sectors (LEK Partnership, 1994, cited in McMahon, 1995) include:

◊ 96 per cent of service businesses are SMEs;

◊ services comprise 20 per cent of Australian exports;

◊ 1 per cent of all Australian service businesses are exporting, but a huge 40 per cent of new entrants now earn most of their revenues from exports.

These findings point to the growing significance of small and medium-sized service businesses in the expansion of exports. Unfortunately, there was no information about the gender of the owners of these service businesses.

A 1996 Austrade survey of over 2 000 companies who used its export services showed that 72 per cent of them were SMEs (reported in ACOB, 1997). The breakdown by industry sector was:

◊ Professional services (12 per cent);

◊ Communications and technology (11 per cent);

◊ Food and beverages (12 per cent);

◊ Rural commodities (9 per cent);

◊ Projects and engineering (14 per cent);

◊ Transport and defence (7 per cent);

◊ Education and training (8 per cent).

It is not known how many of these SMEs are owned by women, nor how representative companies using Austrade's export services are of the general population of exporters.

The Ernst and Young survey of 44 Australian women entrepreneurs found that a healthy 11 per cent of them were engaged in the information technology industry – which suggests that women are moving into non-traditional areas (Ernst and Young, 1996).

Information, advice and business assistance

This section deals with the sources of information, advice and assistance that business women access to establish or maintain their businesses. Specific financial assistance in the form of loans or grants is discussed below.

The need to export has long been recognised by Australian governments as a strategic imperative for all Australian industries. The stimulation of export start-up is seen as a specialised area separate from small business in general. There are now a number of specific government programmes designed to promote and support export with advice and practical assistance.

Underutilisation of assistance

The Australian Government, through its international trade advisory service, Austrade, provides information, advice and assistance to Australian businesses seeking to develop their export capability. While there are no comparative gender data on access rates to export programmes, the general view is that women are underutilising these assistance programmes.

Findings from relevant research

In Australia the use of government sources of assistance by women in business is generally low and lower than the use of independent, non-government sources (The Flinders University, 1995; Tasmanian Women's Consultative Council, 1996).

In their review of research, The Flinders University found that accountants were the most popular source of advice being used by 40 to 80 per cent of women in business. They were followed more distantly by lawyers (The Flinders University, 1995). The Ernst and Young study found that women, in the main, seek advice first from their accountants and then from their colleagues (Ernst and Young, 1996).

Of the ten women interviewed, most had not sought assistance from any government programmes designed to stimulate and support export activities. Six of the ten women (including the three who are not currently exporting) had made no approach to Austrade in the development of their export business and had no experience of using its services. Of the remaining four:

◊ one had used Austrade and state government trade missions extensively to develop markets in Singapore, Malaysia and South Africa;

◊ one had general experience of Austrade services in opening up markets in Europe, the United Kingdom and the United States;

◊ one had tried out Austrade when she was assessing her export potential in Thailand;

◊ one had indirect assistance from the Federal Ministery for Trade and Industry in setting up overseas operations.

Causes of underutilisation of assistance

When it comes to advice and assistance about exporting, women often need the services of specific agencies like Austrade which specialise in export-related issues rather than accountants who can provide general business advice but who may be unfamiliar with international issues.

There may be a number of factors behind the low utilisation of export services. Women, particularly those from other cultures, are not always aware of the assistance that is available in their adopted country. They may prefer to rely on their own knowledge and their personal or business contacts in their home country for advice and assistance.

One migrant woman entrepreneur said she always approached the government of the target country but had never sought assistance within Australia. Another migrant woman, on the other hand, had made extensive use of Austrade services. She, however, had studied and worked in Australia for many years before going into business and this may have familiarised her with the government sector.

The Flinders University review of research noted limitations on the usefulness of business assistance (both government and independent) due to lack of awareness, inaccessibility and lack of relevance to women (The Flinders University, 1995). It recommended further research to determine whether women have special requirements, such as more appropriate timetabling that takes into account the family responsibilities of women. When it comes to seeking information, women may feel intimidated in a male-dominated audience and be more comfortable in an all-female group (ACOB, 1997). Women may also benefit from exposure to female role models (The Flinders University, 1995).

Austrade has recognised this issue and set up the Women in Export programme to address it. A description of its aims and activities follows.

Business assistance best practice example: Austrade's Women in Export programme

The Women in Export programme aims to ensure that women in business are aware of the full range of Austrade's services and have access to them. Other major programme objectives include:

- ◊ encouraging business women and businesses owned or operated by women to export;
- ◊ identifying business women who are active in exporting and who could become role models for other women in business;
- ◊ working with existing women's networks to help them achieve a greater export focus.

Features of the programme to date have been:

- ◊ country-specific seminars, *e.g.* Doing Business in Regional Japan;
- ◊ organising meetings with visiting missions and delegations of business women, *e.g.* business women from Taiwan; Business women's Breakfasts in Melbourne and Sydney;
- ◊ international collaboration, *e.g.* meetings with visitors from the International Women's Federation of Commerce and Industry and joint seminars;
- ◊ joint exhibition with Victorian College of International Business of successful women exporters.

Rather than duplicate the work of existing women's business and professional organisations, the Women in Export programme aims to work with all such groups to facilitate the entry of women into the world of exporting.

The Women in Export initiative has been a success and represents an example of best practice in this area. The Australian Council of Businesswomen Ltd (ACOB) recently endorsed the programme and noted that further expansion of this programme could be a very effective way of encouraging women to launch into international trade or to further develop their international trading and exporting activities (ACOB, 1997).

Rationale and strategies: why and how women go about investing and doing business overseas

Rationale – findings from relevant research

The women who were interviewed for this report were asked why they started doing business or investing overseas. The circumstances that led them to move in this direction were wide-ranging and included:

- ◊ A deliberate strategy to grow an established domestic business by planned expansion into the larger markets overseas.
- ◊ Part of the original business concept – one woman said her business (travel and tourism) was capital-intensive and seasonal, inbound tourism was essential to ensure assets were fully utilised all year round.
- ◊ A deliberate strategy to support the whole business from early on - one woman reported she had marketed into Singapore early in the life of her business to gain credibility in the Australian market.
- ◊ An unexpected opportunity was taken up by one woman entrepreneur – her firm was approached directly and indirectly by private businesses and governments in some countries. In one case, she was approached by her existing overseas distributor to extend her business into another country.
- ◊ A contact in the home country of a migrant woman suggested a business opportunity which became the basis of a new global business based in Australia.

Migrant women or women who have spent a large part of their lives in other countries seem more open to opportunity in the overseas countries they are familiar with. On the other hand, migrant women sometimes lack opportunities in Australia and have to capitalise on knowledge of, and contacts in, their home country as an alternative to domestic business.

By contrast, Australian-born women or migrants of English-speaking background do not have the same familiarity with Asia. This is not a disadvantage – but may mean that migrant women are more likely to see export as an integral part of their business from start-up.

Strategies – findings from interviews with women

The strategies used by the women interviewed were governed largely by the circumstances in which the export opportunity arose and the proven strategies of past business or professional successes. Some examples are given here to illustrate the point:

Example No. 1

One woman had worked for many years as a senior partner in an international accounting and management consulting firm in both Malaysia and Australia, where she had specialised in corporate planning and strategy and had advised businesses on how to establish and grow.

The idea for her international business venture was suggested to her by a former client. She was confident in her ability to establish businesses in an international environment and did not feel the need to seek professional advice or assistance in setting up her business, to do comprehensive research or to prepare a formal business plan. Her own knowledge and experience were sufficient to make the business happen and she financed it from her personal savings.

Example No. 2

One woman had no previous exposure to business but joined her husband's business when he needed help. To help her launch her first export venture she sought the assistance of a government agency to locate a distributor in Singapore. She found that selling into Singapore meant greater credibility in the Australian market and so enhanced her domestic business. Expansion into Malaysia, her home country, came soon after she went on a trade mission to Malaysia organised by her state government.

With this exposure she attracted the attention of JETRO, who approached her about marketing into Japan and Japan is now her biggest overseas market. Her strategy is to develop overseas markets no faster than her resources permit and only after careful preparation. She made good use of government assistance.

Example No. 3

One woman reported that she had used different strategies to expand into Asia depending on the target country, but each new country offered a launching pad for expansion into successive neighbouring countries. For example starting up a network of distributors and support facilities in Singapore made it easier to expand into Malaysia. Similarly starting up in Malaysia facilitated entry into Brunei. Expansion was always carefully planned and ultimately possible only if the skilled human resources were available in the target country.

Example No. 4

One woman determined to go global from the early days of her business. She had previous business experience in a related industry. She marketed her business to agents in target countries to get her products included in distribution and promotion networks. She did this through personal contacts and with some assistance from her Australian industry body. She now has a very successful international business.

Example No. 5

One woman with a successful business in Australia reported that she had tried unsuccessfully to purchase part of a company in Thailand to expand overseas. She found it difficult to get the information she needed to make an informed decision about her proposed investment. In the end she ruled it out as unsound. On the basis of her Thai experience she determined that any future business in Thailand would be developed through a strategic alliance with a local company (*e.g.* a joint venture) and not via direct investment.

Summary of strategies

In summary the strategies women reported using in the interviews were:

◊ joint venture;

◊ informal partnership agreement;

◊ use of a local distributor in the overseas country;

◊ personal contact with businesses in the target market;

◊ showcasing product via government trade missions and exhibitions.

Trade missions: a neglected strategy

Some women reported being involved in trade missions and other government-sponsored international trade activities. This is a strategy that more women could employ. In fact, trade missions need to have a much higher proportion of women on them. There are enough successful Australian business women with an international profile to make this a reality. By excluding women from these activities, we send a message to our international counterparts that Australian women do not do business.

Financing exports

Financing export market development

Doing business overseas is expensive for Australian small and medium-sized enterprises. Exploratory missions to overseas countries are costly. Austrade has encouraged exporters to develop with its Export Market Development Grants which reimburse part of the marketing costs incurred by the business. Two of the women who were interviewed for this report had used these grants to good effect.

However, there are no statistics on the relative numbers of women and men seeking and winning grants. There is already in Australia a requirement on all public and large private sector employers to report annually on progress in addressing gender equity issues in employment as documented in their affirmative action programmes. These reports must include gender-disaggregated data on employment profiles. The numbers of women in employment and their levels of employment are thus monitored through these reports.

When women leave employment and become self-employed they are no longer included in these comprehensive reports. They do count in the Australian Bureau of Statistics, for example, in the Labour Force Survey, but this does not distinguish between employment and self-employment. There is an equal opportunity information gap that includes women who are participating in government programmes directed at the self-employed. In the first instance government agencies should ensure that they keep relevant statistics which will serve to monitor and analyse access to government business programmes by women and men.

External financing of SMEs

Australian research suggests that women are less likely than their male counterparts to seek funds for start-up and growth from financial institutions. The School of Commerce at The Flinders University of South Australia was commissioned by the Australian Government's Department of Industry, Science and Technology to prepare a comprehensive review of Australian and international research into women in small business.

Finance was one of the specific research areas that was reviewed, and findings included:

◊ restricted access to finance is frequently cited as an obstacle facing women in small business;

◊ this is attributed to discrimination by bank officers and disadvantages faced by women such as lack of relevant experience and personal financial resources;

◊ rigorous studies do not show gender as a factor restricting access to finance – access to finance is a problem experienced by all small enterprises;

◊ it is not clear that women are being discriminated against, although many women believe this to be the case (The Flinders University of South Australia, 1995).

While discrimination is illegal in Australia and there is no direct evidence of any discrimination, many women believe that it is more difficult for them to access debt finance from banks and other financial institutions. One woman reported that she gave up dealing with financial institutions once her business was large enough for her to employ men who looked the part and were more appealing to bank officers.

Findings from interviews with women entrepreneurs

All the women interviewed for this report funded their move into export internally. There were no external borrowings. The six women exporters comprise:

◊ four who financed internally from reserves or used personal finances;

◊ one who floated the company as an alternative to debt financing (she said: "I don't like dealing with banks");

◊ one who divested her domestic business to finance an international business after banks refused to give her a loan.

Possible solutions: special financing arrangements through banks

Perhaps Australia needs something similar to the initiative of the Wells Fargo Bank, which announced in 1995 that it would open a US$1 billion credit line for women who had run profitable businesses for at least two years. The Bank's intention was to give recognition to what was an important resource within the US economy – women in businesses. A consortium of Australian banks could create a similar facility. However, it is not always the case that women need a bank of last resort as the following example illustrates:

A loan guarantee scheme was implemented by the Victoria Women's Trust for women on low incomes wishing to establish a small business who were unattractive to banks and who could be financed by the State Bank of Victoria with the Women's Trust acting as guarantor for the loan. Interestingly, the criteria for acceptance included the preparation of a high-quality business plan backed by appropriate market research. It was found that once the criteria were met the proposal was sufficiently attractive to banks without the loan guarantee.

The loan guarantee programme has been terminated and the Trust now focuses on providing assistance to women wishing to develop business plans (The Flinders University, 1995). This suggests that what women need (rather than government financing interventions) is help in developing sound business plans, and someone to put them in touch with investors who are keen to lend to small businesses owned by women.

Possible solutions: business networks and mentors

The Flinders University review of research reported (The Flinders University, 1995):

- ◊ networks are identified as a key strategy to secure access to finance and business partners;
- ◊ women face difficulties gaining access to business partners as a source of additional resources; a register of proprietors seeking a partner may serve to facilitate access to such resources.

The Flinders University examined a review of the literature by Borzi which noted:

- ◊ male business owners typically draw their partners from active networks dominated by business and male friends who also happen to be those with greater wealth and access to finance;
- ◊ women's active networks did not tend to include such people but rather women partners and close friends;
- ◊ women are disadvantaged in their prospects of finding investors, business partners and commercial information because of these network characteristics (Borzi, 1994, reported in The Flinders University, 1995).

They also noted a study of Chinese and Indian businesses in Australia which highlighted the importance of women's kinship networks in providing skills and resources to family enterprises (Ip and Lever-Tracy, 1992, reported in The Flinders University, 1995).

The Flinders University review concluded that the establishment and extension of business networks and mentoring programmes which specifically address the needs of women proprietors should be supported. It further concluded that particular networks supporting cultural minority business operators should also attract policy attention.

The overall picture is not clear. The issue of access may be confused with the gender-based differences in the circumstances of those presenting to banks for finance. While more research may unravel the issue, effective measures are needed now to facilitate small business borrowings by both male and female entrepreneurs. However, we should not lose sight of the fact that there is a real perception by women that they are disadvantaged. The following section gives an example of a successful mentoring programme which is operating in one state in Australia; namely, New South Wales.

Mentoring best practice example: the Step Ahead programme

Step Ahead is part of the Women in Business Mentoring Programme which began operating in NSW with a pilot in 1995. It provides opportunities for women in business to learn from each other (NSW Department of State and Regional Development, 1996).

The objectives of the programme are:

- ◊ to establish a pool of women to act as role models for other business women;
- ◊ to enhance the growth and viability of businesses owned by women leading to wealth creation for women and job creation for the State of NSW.

The target group is female, small business owners who have been in business for 18 months or more and are looking to expand or export. The programme consists of 50 hours of contact comprising:

◊ 10 hours of business roundtables;

◊ 20 hours of business skills training;

◊ 20 hours of one-to-one mentoring.

The training covers business planning, including export options, marketing, human resources, management and conflict resolution, finance and administration.

Results of the 1995 pilot mentoring in the participants' businesses ranged widely and included:

◊ 74 per cent of participants increased their turnover during the six-month period;

◊ 52 per cent increased the number of employees in their business;

◊ 60 per cent increased return on investment;

◊ 70 per cent increased market share;

◊ almost all reported increase in personal and business skills and business confidence.

The pilot Women in Business Mentoring programme has proven successful and this promises well for *Step Ahead*. There is no reason why the programme should be confined to New South Wales.

Assets that women bring to the global market

Findings from interviews with women entrepreneurs

The women interviewed for this report believed they have particular skills that advantage them in their dealings with people in business or government in other countries:

◊ They are especially skilled in building long-term personal relationships with the people they seek to engage in business. Women are patient and this is important in Asia. In Asia, business is personal and long-term.

◊ They are good at building and maintaining business networks from small beginnings. This often has to be done when opening up a business in a foreign country. Men use business networks, but frequently they were born into them.

◊ Part of relationship-building is effective communication skills. Women are excellent communicators and communication is especially important when doing business with people from another culture and/or language.

◊ Women are sensitive to cultural differences and the importance of behaving appropriately when visiting other countries (this is not surprising, after all women have been doing business in the male-dominated world of Australian business for a long time. To Australian women that is just another culture.)

◊ Women are good organisers and pay attention to detail. This is especially important when operating for the first time in a very unfamiliar environment such as a foreign country. Men can be over confident and let details slide.

In some circumstances women have particular advantages. They are non-threatening, often because they are not taken seriously but also because they are less aggressive. Asian societies appreciate a more gentle, less aggressive approach. One woman interviewed for this report who was involved in leasing and buying aircraft for her business said she was a novelty in this masculine industry and therefore made more of an impact.

Migrant businesswomen and their overseas networks

There is no research data on migrant businesswomen's overseas networks. However, it is worth noting that across much of Asia, prominent ethnic Chinese business people in wholesale and retail trade control a web of distribution networks (East Asia Analytical Unit, Department of Foreign Affairs and Trade, 1995). They are also significant investors in banking and finance, and in manufacturing industries. There are over 55 million overseas ethnic Chinese, 50 million are in Asia, 3.4 million in the Americas and 600 000 in Oceania (including Australia) (Ch'ng, D., 1993, reported in the East Asia Analytical Unit, Department of Foreign Affairs and Trade, 1995).

Interviews for this report with migrant businesswomen show that they have strong Asian networks and are able to use them to their advantage. There are opportunities for other business women to tap into this network by working with these migrant business women.

The barriers faced by women doing business overseas

Cultural differences in attitudes to women

For women in Australia today there are still some barriers to doing business overseas. In Japan (and similarly in the Republic of Korea) it remains the case that women in senior ranks in business are a rarity. When an Australian woman goes to Japan to do business, she is automatically regarded as the junior partner. Any man who accompanies her is assumed to be her senior. Australian women sense that Japanese businessmen feel uncomfortable dealing solely with them, not least because they present one very difficult problem. How can they be taken to the traditional all-male entertainments which are such an important part of doing business in Japan?

This is not the case in South-East Asia, India or in China. There are many successful business women in South-East Asia and many senior ranking women in government in China. But in Japan this is not so, and women need to be very aware of these culturally derived gender differences. They also need to be alert to cultural expectations of behaviour. They need to dress, speak and behave appropriately. This is true of men also, but women are less easily forgiven for any transgressions.

Of course, Australian women (regardless of their country of origin) experience these gender differences as much at home in Australia as they do overseas. That is, their male counterparts or the male employees of institutions they deal with, treat them differently. Some women find that the effort of overcoming the prejudice against them is so great that they simply delegate the task to suitable male employees or partners.

Family responsibilities

Lastly, it is still the case that women in business are constrained by their family responsibilities as they strive to integrate their family life and their business aspirations. Women seek balance in their lives. This is confirmed in Australian research and seems to be true for women all over the world. For Australian women wanting to break into overseas markets there are special problems. For example, international travel, which is so essential to developing good personal relationships, is more time-consuming from an Australian base, and the conflict between family and business is correspondingly greater.

The women who were interviewed saw no special disadvantages in being a woman. Many pointed out that women are treated differently overseas, but the same is true in Australia. In fact, some said they were treated better by overseas males than by their Australian counterparts (and some said they were better received by women overseas than by women in Australia). At the end of the day, they all agreed that if you were good at your business, then there were few insurmountable barriers to success. *Business is business whether it is in your own country or somewhere else, for both women and men alike.*

This does not mean it is easy for women. Many women believe they have to be more careful. When women make mistakes they are not easily forgiven. Women are more easily discounted in business. Women have fewer options for the obligatory cementing of business relationships in social settings – on the golf course or in the nightclub. They can be disadvantaged by the golf-course factor and have to work creatively to overcome it.

Policy initiatives

Findings from interviews with women entrepreneurs

Women entrepreneurs interviewed made many suggestions on how to improve the business environment in which they operate. This is a summary of the feedback they gave about the situation of women:

◊ greater recognition back home of the role of migrant and Australian-born women in international trade;

◊ more forums where women can showcase their activities, *e.g.* international conferences for business women, exhibitions;

◊ mandatory inclusion of women in trade missions and other government trade initiatives and in policy- and decision-making bodies;

◊ formal recognition through the taxation system of the child care and other family responsibilities that women have to juggle with their business activities;

◊ women doing business in Japan need support from Austrade; perhaps Austrade could provide formal introductions to ensure that their rank is clearly recognised.

Some women also commented that single-sex high schools with strong business skills development in the curriculum would advantage women. Others felt that business women needed more information about the importance of international markets and awareness of export incentives right from the start-up of their businesses.

Lack of political influence

Women everywhere comment on their lack of political influence. Many seem to feel the lack of well-developed networks and mentors. The Flinders University review of research found:

◊ "women business owners are less prevalent than men in positions of influence in economic development organisations and other bodies such as local, state and federal government policy and planning groups";

◊ "they need to be included in regional economic planning processes and at all levels of government policy and planning affecting business" (The Flinders University, 1995).

There are many business organisations to which women can belong which provide personal support, social status, influence, information, advice and assistance and opportunities to make contact with potential business partners. Few of them focus on the strategic issues which impact on women in business. Few of them provide an effective channel for women's views into the offices of government policy and decision makers. A newly formed Australian body called the Australian Council of Businesswomen Ltd aims to do just that.

Lobby best practice example: The Australian Council of Businesswomen Ltd (ACOB)

This organisation was founded in 1995 with a single mission: to ensure that the opinions, ideas and vision of Australia's business women form part of the top-level debate which influences the creation of both wealth and employment in Australia (ACOB, 1996). ACOB serves as a peak body for all business women – from micro to mega businesses – delivering their views direct to decision makers.

The Australian Government provided seed funding to establish ACOB in November 1995 in recognition of the need to consult with business women. The proliferation of organisations, networks and groups made this difficult, and the government recognised the immense value of a single peak body focused on providing a quick and well-informed response on any issue affecting business women in Australia.

ACOB aims to:

◊ develop, comment on and lobby on policy and initiatives;

◊ develop and disseminate relevant information and research;

◊ promote achievements of women in business;

◊ participate in international networks.

To date ACOB has made substantial progress:

◊ it has organised business women's summits to consult with women and government;

◊ government has sought advice on policy matters from ACOB;

◊ ACOB has signed a Memorandum of Understanding with the Ladies' Wing of the Indian Merchants' Chamber – an example of collaboration among national bodies to achieve a greater international profile for business women.

It is hoped that ACOB can give business women the political access they need and which would reflect their growing importance.

Conclusions and recommendations

Gender-specific statistics

Australian business women in SMEs are investing and doing business overseas. However, there are no gender-specific statistics available on the nature and extent of these activities. The lack of this kind of information hampers the development of sound policies to encourage business women to export and makes it difficult to monitor the effectiveness of any initiatives that are implemented.

Recommendation 1: Existing data collections held by the Australian Bureau of Statistics should be modified to permit the gender breakdown of industry statistics relevant to SMEs involved in export and other international business.

Business assistance programmes: special needs of women

The use of government sources of assistance by women in business is generally low and it is likely to be low for women business owners who are involved in exporting. This may be because women are unaware of these services, or because they are inaccessible or not relevant. Women may need more appropriate timetabling. They may also be more comfortable in an all-female environment. Austrade has had some success with its Women in Export programme which tries to ensure that women are aware of its services and have access to them.

> ***Recommendation 2:*** In designing, marketing and delivering export advice and business assistance programmes, agencies should be aware that many women may have special needs.
>
> ***Recommendation 3:*** Austrade should be encouraged to maintain and develop its Women in Export programme to meet the needs of women.

Gender specific statistics: export assistance

Women are making use of the Australian Government's Export Market Development Grant scheme although some women reported their ventures were too small to qualify for the grant. However, no direct information about the comparative application and success rates of women and men for these grants exists.

> ***Recommendation 4:*** Gender-based statistics should be maintained by Austrade and other relevant government agencies which permit the analysis of comparative application and success rates by women and men for export assistance including Export Market Development Grants.

Financing exports by women

Women in general are less likely to seek funds from banks and other external sources for start-up and growth of their business ventures. Interviews were carried out for this report with Australian business women who are involved in overseas business. The results suggest the same is likely to be true in the particular case of export start-up and expansion. Women have a perception that banks discriminate, but there is no rigorous research to confirm this perception. Women also face difficulties in finding business partners.

> ***Recommendation 5:*** Banks and financial institutions should design products and services that are women-friendly and can be used by them to finance their export activities.

Mentoring programmes

Mentoring programmes may assist women in obtaining business partners and dealing with financing issues. One such programme directed at women in small businesses wishing to export is *Step Ahead.*

> ***Recommendation 6:*** The government should continue to support the business mentoring programme *Step Ahead* and to expand it nationally. More mentors, both men and women, should be encouraged to participate in the programme.

Gender balance on official government trade missions and other government overseas business programmes

Women are using a variety of strategies to open up markets overseas. However, few of them participate in official trade missions and organised exhibitions.

Recommendation 7: Federal and state governments should strive to improve the gender balance in the composition of government-funded trade missions. All official trade missions organised by the government should comprise at least one-third women.

Acknowledgements

I would like to thank those women entrepreneurs who agreed to be interviewed for this report. Their experiences and comments helped me formulate my thinking for this paper.

I would also like to thank the Australian Council of Businesswomen Ltd., Mr Trevor Tappenden, National Director, Entrepreneurial Services, Ernst & Young, and the many well wishers for their support and encouragement.

To Gayle, Ramona, Anne-Marie and my son, Kenny, a special thank you.

BIBLIOGRAPHY

AUSTRALIAN BUREAU OF STATISTICS (1993), *Small Business in Australia,* ABS Catalogue No. 1321.0, Canberra.

AUSTRALIAN COUNCIL OF BUSINESSWOMEN (1996), *If You're a Woman and You're in Business, You **Need** to Know about ... ACOB.*

AUSTRALIAN COUNCIL OF BUSINESSWOMEN (1997), "Women Entrepreneurs – Globalisation and International Trade", internal report.

BORZI, A.M. (1994), *The Gender Finance Gap*, Smythe Pty Ltd, Sydney.

EAST ASIA ANALYTICAL UNIT (1995), *Overseas Chinese Business Networks in Asia,* Department of Foreign Affairs and Trade, Canberra.

ERNST AND YOUNG (1996), *Australian Entrepreneurial Survey*, Second Annual Survey of Australian Entrepreneurs: Thought Leadership Series.

IP, D. and C. LEVER-TRACY (1991), "Asian Women in Business" in *Women in Migration*, Bureau of Immigration Research, Melbourne.

INDUSTRY COMMISSION/DEPARTMENT OF INDUSTRY, SCIENCE AND TOURISM (1997), *A Portrait of Australian Business,* Results of the 1995 Business Longitudinal Survey, ACT.

KERMOND, C.L., K.E. LUSCOMBE, K.W. STRAHAN and A.J. WILLIAMS (1991), *Immigrant Women Entrepreneurs in Australia*, Working Papers on Multiculturalism No 14, University of Wollongong, Wollongong

LEK PARTNERSHIP (1994), *Intelligent Exports and the Silent Revolution in Services,* Australian Trade Commission, Sydney.

MCKINSEY AND COMPANY (1993), *Emerging Exporters: Australia's High Value-added Manufacturing Exporters,* report prepared for the Australian Manufacturing Council, Melbourne.

McMAHON, R.G.P. (1995), *The Internationalisation of Small and Medium Sized Enterprises: An Australian Perspective,* in March Briefing 1995 of the *Asian Business Supplement,* South Australian Centre for Economic Studies, Adelaide and Flinders Universities.

McMAHON, R.G.P. (1997), Private communication, School of Commerce, The Flinders University.

NATIONAL BOARD OF EMPLOYMENT, EDUCATION AND TRAINING (1994), *Making it Work: Women and Small Business Small Business,* Employment and Skills Supplementary Report, Commonwealth of Australia, Canberra.

NSW DEPARTMENT OF STATE AND REGIONAL DEVELOPMENT (1996), *Women in Business Mentor Program: Information Booklet.*

TASMANIAN WOMEN'S CONSULTATIVE COUNCIL (1996), *Women and Small Business: A Tasmanian Perspective,* report to the Minister for the Status of Women.

THE FLINDERS UNIVERSITY OF SOUTH AUSTRALIA, DEPARTMENT OF COMMERCE (1995), *Women in Small Business – A Review of Research*, report prepared for the Department of Industry, Science and Technology.

Chapter 7

A RESOURCE PERSPECTIVE ON WOMEN'S ENTREPRENEURSHIP: RESEARCH, RELEVANCE AND RECOGNITION

by
Candida G. Brush
Assistant Professor, Management Policy
Boston University, Boston, United States

Women entrepreneurs come to venture creation with different endowments of human and social capital than their male counterparts. These differences are explained generally by the pervasive influence of similar social structures – work, family and organised social life – across all OECD countries. Social structures differentially affect women's human and social capital endowments, leading to unique combinations of organisational resources in their businesses. While many of the endowments that women entrepreneurs bring to their entrepreneurial endeavours can be assets in their new organisations, improved academic research, relevant educational materials and visibility of successful women entrepreneurs could significantly improve resource access and fill gaps. In particular, for the continued success of women's entrepreneurship into the 21st century, public policy needs to pay greater attention to research, relevant education and recognition.

The growth and presence of women entrepreneurs in SMEs is increasingly important in the global economy. Statistics show that the numbers of women business owners rose during the last decade in nearly every OECD country. For example, the United Kingdom reported an average 10 per cent growth every year between 1981-86. The Netherlands showed a gain of 36 per cent between 1975-85 (OECD, 1990). Spain, Sweden, Canada and the United States also show increases in the numbers of women-owned businesses, ranging between 10-20 per cent. While women still have a smaller share than their male counterparts, the share of women's business ownership ranges from 37 per cent in some countries (the United States) to 30-35 per cent in others (Australia, Canada, the Netherlands, Japan) to 20-25 per cent in yet others (Spain, the United Kingdom, Germany) (OECD, 1990; NFWBO, 1996; Institute for Small Business, 1996).

Concurrent with this growth in numbers of women business owners are their greater contributions to country economies. In the United States more than 7.9 million women-owned businesses add more than $250 billion to the US economy and provide more than 11 million jobs. Total revenues of these businesses tops $2.3 trillion (NFWBO, 1996). In Canada, there are more than 700 000 women-led firms, providing 1.7 million jobs (Institute for Small Business, 1996), while in Australia, women-owned businesses account for about 10-15 per cent of GDP (The Flinders University, 1995).

The importance of women entrepreneurs in OECD countries draws attention to the need for a better understanding of the approaches women take to create and grow their own enterprises. Unfortunately, theoretical, empirical and educational advances have not kept pace with the growth in numbers and presence of women entrepreneurs. For instance, research focusing on or including women number less than 10 per cent of all academic investigations (Baker, Aldrich and Liou, 1997). Another review of the Proceedings of the Babson College Research Conference on Entrepreneurship, *Frontiers of Entrepreneurship,*[1] categorised 227 studies published between 1980-87 and found that only 13 (6 per cent) investigated women and minorities (Churchill and Hornaday, 1987). Between 1988 and 1993, this author identified a total of 522 studies and summaries published in this Proceedings. Of these, only 17 (3 per cent) focused on women-owned business and 35 (7 per cent) included gender as a variable, totalling about 10 per cent of all studies. This percentage is actually high, compared to publication records of other entrepreneurship journals, which Liou and Aldrich (1996) found to be about 7 per cent.

As a consequence of the proportionately smaller number of studies focusing on or including women entrepreneurs, less is known about the ways in which women develop, manage and grow their ventures. Literature reviews show that more than 50 per cent of all studies focus on personal characteristics of women entrepreneurs (Brush, 1992; 1996), while the remaining studies cover a wide array of topics including: the start-up process, environmental factors and organisational strategies/policies. Of greater concern is the lack of theory employed in studying women entrepreneurs, less than 30 per cent of all studies, and of those employing theories, many are based on theories emerging from studies of men. Because women are not included or, if they are included, are studied as part of the entrepreneurial phenomenon, the extent to which women create different organisations, or manage differently from men is not well understood (Brush, 1996).

In sum, the growth in numbers and contributions of women entrepreneurs is outpacing academic research and theory development. This is an important gap. To better understand the phenomenon of entrepreneurship, populations of women need to be included or studied separately in both academic and government investigations and this research must be grounded in theory or conceptual frameworks. From this a better understanding of the similarities and differences between women and men entrepreneurs will emerge, as will areas where public policy assistance is needed. The remainder of this paper develops a conceptual model of venture creation using a resource-based view of organisations. Then, causes of potential differences based on gender are discussed and organisational dimensions (structure, policies, strategies, values and practices) characteristic of women-owned businesses are proposed. This paper ends with recommendations for public policy attention that emerge from this discussion.

New venture creation process

Resources are the foundation for all organisations (Penrose, 1959; Andrews, 1971). Without resources, an organisation cannot produce products/services, develop unique advantages or design and implement strategies (Wernerfelt, 1984). Emerging theory from strategic management, referred to as the resource-based view (RBV), argues that because resource markets are imperfect, firms will assemble and create heterogeneous resource bundles that define firm differences, determine the competitive space each can occupy, and shape perceptions of future opportunities (Wernerfelt, 1984; Rumelt, 1987). Resources, or *stocks of available factors* represent the basic unit of analysis in the productive process (Amit and Shoemaker, 1993, p. 35). Over time, resources can evolve into capabilities or core competencies – those activities essential to the firm's mission that it executes consistently well (Prahalad and Hamel, 1990). Resource combinations that are valuable, rare, inimitable and non-substitutable are those most likely to yield a distinct competitive advantage in the market-place (Collis and Montgomery, 1995). RBV theory emphasizes attention to the future, and development of "invisible assets" that optimally lead to a "sustainable competitive advantage" (Itami, 1987). The connection between resource choices and competitive advantage is highly dependent on the types of resources owned or controlled by a firm, their combinations and their dimensions (Penrose, 1959; Hofer and Schendel, 1978; Barney, 1991). For the most part, this theory is intended to apply to established organisations.

New ventures face different resource conditions than established organisations. The process of new venture creation is the "organising (in the Weickian sense) of new organisations" (Gartner, 1985, p. 697). This process includes: *i)* the recognition of the opportunity; *ii)* the definition of a project or plan to exploit that opportunity; *iii)* the mobilisation of resources necessary to put the plan into action; *iv)* the intention to create value; *v)* the appropriate management of risk (Bygrave and Hofer, 1991). However, in contrast to established organisations, the resource base is evolving. At the most basic level, the new venture is "resource-free " in that the organisation has no resources until the entrepreneur identifies the needed resources, and commits these to a particular opportunity or strategic purpose following her vision of the new organisation (Greene, Brush and Hart, 1997). As resources are committed to an opportunity, resource partners are engaged and resources are combined. The "organising" process yields a new organisation. But, if the entrepreneur fails to acquire resources, or if she acquires resources that are insufficient or inappropriate, the venture is unlikely to survive its infancy.

While the organisation may be "resource-free" at the start, the entrepreneur starts with a set of endowments, human and social capital. Human capital includes the achieved attributes, education, experience and reputation of the entrepreneur (Becker, 1964; Cooper, 1981); while social capital is comprised of the relationships, networks, family, race, ethnicity and political connections of the entrepreneur (Bordieu, 1983; Glade, 1967). These "endowments" provide the basis from which the entrepreneur will acquire physical resources (tangible assets, facilities, technology), and financial resources (funds), leading to the formation of a new organisation (Greene, Brush and Hart, 1997). Organisational resources are comprised of the structures, policies, routines and procedures, culture, values, knowledge and organisational alliances (Tomer, 1987; Hofer and Schendel, 1978). In the early stages, the entrepreneur's resources and those of the venture may be identical, but a value-creating enterprise will soon need a broader resource base, expanding beyond the founder's resources. The entrepreneur will "leverage" her endowments to obtain access, control or ownership of the financial, physical and organisational resources needed to develop the venture according to her vision of the business (Brush and Bird, 1996). This process is illustrated in Figure 1.

Figure 1. Resources in the new venture creation process

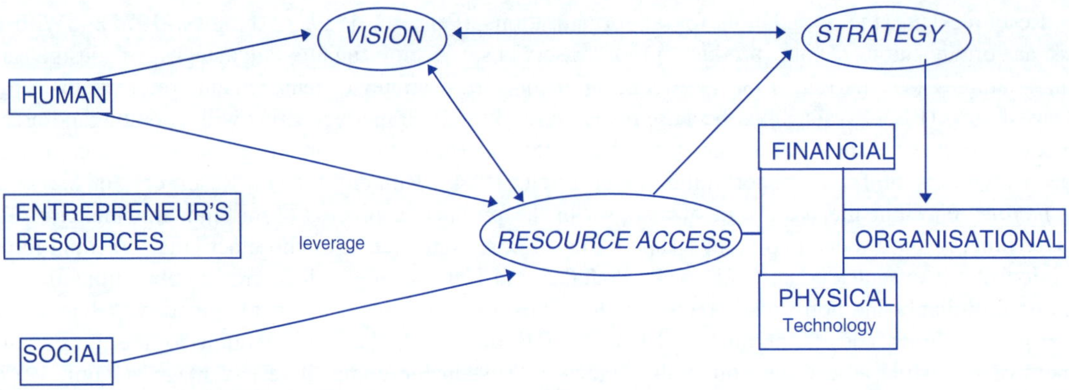

Gender influences on the venture creation process

Following this conceptual framework, it is clear that each new venture is unique, in the starting *endowments* of the founder, the vision and the combination of resources that are brought together to comprise the new organisation (Gartner, 1985). This generalised model details the process of venture creation, showing the crucial importance of human and social capital in this process. Extensive research examines individual motives for entrepreneurship showing that "pushes" and "pulls" influence venture start-up (Brush, 1992). However, gender differences based on motives or individual characteristics are not conclusive. More specifically, psychological motives, such as a desire for independence or achievement, may combine with economic necessity or desire for wealth to motivate venture creation in both men and women. On the other hand, other studies show that men may be motivated by a desire to "be an entrepreneur" or not work for someone else, whereas women may wish to have "flexibility" in balancing work and family or to "help others" (Thompson and Hood, 1991; Chaganti, 1986; Holmquist and Sundin, 1988).

This individual-level research focus has led to an examination of the factors affecting personal career paths, less often including relevant aspects of organisations (Izraeli and Adler, 1994; Brush, 1992; Brush, 1997). Shifting the level of analysis to the environmental context or social structures that influence the development of human and social capital yields a more fruitful basis for understanding gender differences. There is evidence that the social structures (work, family and organised social life) create different circumstances for women in OECD countries (Lerner, Brush and Hisrich, 1997). These social structures have a fundamental impact on the endowments of human and social capital that women entrepreneurs bring to their businesses at start-up.

Social structures vary widely in developed and developing countries (Truman and Allen, 1993). The influences of social structures (work, family and organised social life) affect women's access to entrepreneurial opportunities (Aldrich, 1989). In spite of recent public policy changes in all OECD countries over the last decade intended to increase women's participation in all industrial sectors, occupational. segregation and under-representation in upper-level management positions continue to persist. For instance, in Israel, almost three-quarters of the female labour force are employed in only three of nine major occupational categories, while half are concentrated in only eight of 90 occupations (Izraeli, 1994). In Sweden, 90 per cent of jobs as typists, nurses and housekeepers are filled by women, while 60 per cent of women work in the public sector compared to 23 per cent of all men (NUTEK, 1996). Similarly, in Australia, Finland and Denmark the labour market remains gender-segregated and equal opportunity has not been achieved (The Flinders University, 1995; Hanninen-Salmelin and Petajaniemi, 1994). Both east and west Germany are characterised by gender-specific segmentation of the labour market, with nearly 70 per cent of working women concentrated in ten of 400 occupational categories in west Germany. In east Germany women comprise 92 per cent of those employed in social services, and more than 75 per cent of those employed in education, retailing, and health care (Antal and Krebsbach-Gnath, 1994).

Besides occupational segregation by sector, women also face barriers in rising to executive-level management positions. In Japan and Korea, women are seldom considered for high-level executive positions (Steinhoff and Tanaka, 1994), and in France, women in corporate management positions comprise 13.6 per cent of the total (Serdjenian, 1994). Even in the United States, while 25 per cent of all managers of *Fortune 200* companies are women *(Wall Street Journal,* 1994), the range for some sectors is as low as 7 per cent with a high of 40 per cent in others. Studies in the United Kingdom report that only 10 per cent of the 200 largest UK industrial companies have women board members (Hammond and Holton, 1994). According to a United Nation's report (1991, p. 82) in almost every country the workplace is segregated by sex.

Structural environmental conditions are evident in the workplace, but also in family life. Expectations about the family roles of women restrict women in terms of time, acceptance as business owners and child/dependent care responsibilities. Most women in OECD countries are expected to manage the household and assume a primary role with the family. In Sweden, women do not receive the same level of support in entrepreneurship from other family members as do their male counterparts (Johannison, 1996). In Australia, family issues and child-care concerns are different for men and women, and women are more likely to perform domestic maintenance and caregiving tasks regardless of their involvement in business activities (The Flinders University, 1995). Young, educated, single Japanese women are hired with the expectation that they will "retire" into marriage and motherhood within a few years. After having children, they may return to the workforce on a part-time or full-time basis, but cultural norms of lifetime employment and seniority make it difficult for women to have significant management careers (Steinhoff and Tanaka, 1994). In west Germany, nearly 60 per cent of 25-35 year old women leave the labour force to have children (Antal and Krebsbach-Gnath, 1994). Even in the United States, the burden of child and dependent care falls on the shoulders of women who are primarily responsible as caregivers and running the household (Aldrich, 1989; Powell, 1993; Brush, 1997). A United Nations Report (1991, p. 83) states that "women everywhere retain the primary responsibility for unpaid housework".

In addition to similar workplace and family structures that affect women and men in different ways across OECD countries, organised social life also varies. In Australia, women's access to potential investors or business partners via networks is conditioned by their comfort level with "anglo-masculine" modes of interaction (*i.e.* alcohol and sport), as well as gendered perceptions of their capacities (The Flinders University, 1995). Networks for women in the United Kingdom are often formalised and sex-based, with statistics showing that women are less likely than men to attend conferences, represent local union branches or sit on working committees (Hammond and Holton, 1994). Evening socialising in Japan in restaurants and bars following the workday remains heavily male-oriented, although increasingly women are able to participate as part of the group (Steinhoff and Tanaka, 1994). Comparing women from Italy and the United States, one study found that women tend to have a larger number and higher diversity of kin ties in personal networks (Aldrich, 1989). Another study from Canada finds that women are more likely to name their spouses first in their networks (23 per cent), while only 9 per cent of men respond similarly (Staber, 1996).

Hence social structures-work, family and organised social life consistently differ for women and men across OECD countries. Clearly the extent and combination of these influences will vary by country and life situation of the woman entrepreneur. However, these social structures directly influence the endowments of human and social capital of women entrepreneurs, as well access to financial and physical resources, and ultimately, the organisational resources of the new venture. Figure 2 shows the influence of social structures on the resource model of new venture creation.

The effects of differential work experiences by industry affect the range of opportunities for entrepreneurial endeavours. Research shows that experience in the area of the industry is one of the primary determinants of success for women entrepreneurs (Brush, 1992; Cooper and Gascon, 1992). Hence, women will have fewer human capital endowments in some industries (*i.e.* manufacturing, agriculture, mining, utilities), and more in others (services, communication, personnel and education). Similarly, fewer chances to advance to top-level administrative and managerial positions limit the competencies and decision-making experience women are able to develop at the executive level in for profit organisations, but, on the other hand, they may have more experience in administrative or public services.

Figure 2. Influences of social structures on resources in the new venture creation process

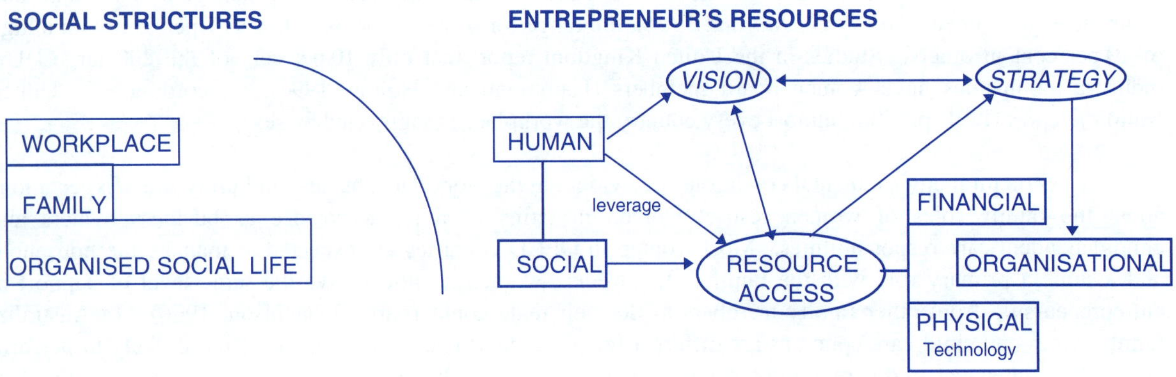

The expectations and role of the family means that women may have less opportunity to devote full-time to work endeavours, being subject to time fragmentation, career interruption and, frequently, lower expectations for success in entrepreneurial endeavours (Brush, 1997). Yet, women often become accustomed to flexibility, to managing multiple tasks and interruptions as a result of their child-raising responsibilities. Likewise, women may be left out of some networks leading to information or expertise due to their interrupted career paths or family responsibilities. Or, they may develop strong social skills leading to unique mutual relationships, trust and commitment that the entrepreneur nurtures with body and soul to make the enterprise become a reality (Johannison, 1994).

In sum, there is evidence across all OECD countries that women encounter social structures in work, family and social life that influence development of human and social capital, different from their male counterparts. The success of the new venture creation process hinges on the ability of the entrepreneur to leverage her social and human capital to create a new organisation. From this, it follows that the organisational resources comprising ventures created by women entrepreneurs will be different than for men. Recognising that each organisation is unique in its bundle of resources (Penrose, 1959), it is nevertheless possible to speculate that the organisational resources of women-owned businesses will differ from previously recognised dimensions which emerge from male-founded organisations.

Organisational resources in women-owned businesses

Organisational resources are comprised of the structures, policies, routines and procedures, culture, values, knowledge and organisational alliances (Tomer, 1987; Hofer and Schendel, 1978). These dimensions are summarised in Table 1.

The organisational resources characteristic of traditional male-founded organisations are well documented (Hofer and Schendel, 1978). Increasingly, academic studies show that women-owned businesses exemplify different organisational resource characteristics. These variations in organisational resources in and of themselves do not necessarily disadvantage women entrepreneurs. In fact, while comparative studies of male- and female-owned businesses are relatively sparse, several studies reflect that women-owned businesses do not perform differently than male-owned businesses in terms of employment growth, survival or profitability

(Johnson and Storey, 1993; Kalleberg and Leicht, 1991; Cooper *et al.*, 1994; NFWBO, 1996; Institute for Small Business, 1996).

Table 1. Organisational resources

Resources	Traditional	Women-owned businesses
Structure	Hierarchy	Horizontal, network *(Brush, 1992; Chaganti, 1986,- Adler, 1994,- Westerberg, 1996)*
Policies/practices	Economic Instrumental	Nurturing, relational, social *(Brush and Bird, 1996,- Holmquist and Sundin, 1990)*
Decision-making	Analytical Fact based	Intuitive, participative *(Chaganti, 1986,- Westerberg, 1996)*
Values/assumptions	Competition Transactional Homogeneity	Integration of family, work, society Reciprocity Diversity *(Holmquist and Sundin, 1990,- Thompson and Hood, 1991; Westerberg, 1996,- Vokins, 1993)*

Other studies show that women-owned businesses grow at a slower rate or may be smaller overall than men-owed businesses (Equal Opportunities Commission, 1996; Kalleberg and Leicht, 1991). Hence the ultimate outcome, as measured in economic terms, may not vary significantly between men and women business owners – although caution is needed in this generalisation because the few studies conducted in this area do not even go as far as breaking down businesses by sector or age, or else tend to consider other outcomes, such as personal satisfaction and organisational effectiveness (Brush, 1992).

What appears to differ is the way that women architect or construct their ventures, and the resource combinations of these organisations. These differences from male-founded organisations, present both opportunities and challenges for women entrepreneurs into the 21st century. As noted above, the endowments of human and social capital may in some ways be advantages for women creating organisations in the service sector, or in designing flexible home-based organisations that take advantage of new telecommunication and computer technologies. Furthermore, predictions of increased globalisation, greater workforce diversity and a need to have boundaryless organisations appear to more closely fit the organisational resources characteristic of women-owned businesses. Indeed, many women entrepreneurs believe they will have unique advantages because of their implementation of social policies such as day-care and family leave, in addition to their willingness to embrace a racially and ethnically diverse workforce (Brush, 1997). However, in other areas, there is a need for changes in public policy across all OECD countries to enhance the possibilities for women to create new ventures. The pervasive influence of differential social structures suggests particular issues needing public policy attention.

Research, relevance, and recognition: issues for public policy attention

Research

As noted in the introduction to this paper, the amount of research including or focusing on women entrepreneurs is not parallel to their growth in numbers or contributions to the economies of all OECD countries. As noted earlier, scholarly research on women business owners represents only about 10 percent of all entrepreneurial research (Baker *et al.*, 1997). Likewise, of more than 250 studies of small businesses funded by the US Small Business Administration (SBA) since 1970, fewer than 20 have included or focused on women. Only recently has the SBA begun to collect and analyse data on women-owned businesses (US Government, 1993; 1995). Furthermore, the governments of Canada, Australia and Sweden have recently commissioned studies to identify the numbers, sectors and participation of women entrepreneurs in these countries (Institute for Small Business, 1996; NUTEK, 1996; Equal Opportunities Commission, 1996).

Yet, the general lack of inclusion of women in academic and government research studies, means that we understand less about entrepreneurship in general. Our theories and prescriptions for success are based largely on research and findings about male entrepreneurs. We know little about the ways in which women architect their new ventures, identify, acquire and combine resources to create new organisations. Better quality research that identifies the "resource gaps" that women may have will permit more effective targeting of programmes and assistance. Public policy makers should consider the following:

1. Provide funding for scholarly research that compares the venture creation process and resource assembly of male and female-owned businesses.

2. Ensure that women-owned businesses are included in all government studies, and that sex identification is standardised on all forms relating to small and medium-sized business activities.

Relevant education

In addition to the gap in research, educational materials are also deficient. While women's access to entrepreneurial education, training programmes and assistance has improved dramatically over the past few decades (OECD, 1990), key decision makers featured in stories, textbooks and cases of successful entrepreneurs are, in the vast majority, male. The best known source of business school cases, the *Harvard Business School Case Catalogue,* features women in fewer than 10 per cent of its entrepreneurial cases. Recognition of this has motivated women's organisations such as the National Foundation of Women Business Owners (NFWBO) to write cases featuring women as key decision makers. The venture creation process reflecting the human and social capital endowments of women and their unique organisations needs to be included in text books, cases and seminars on entrepreneurial education. By omitting success stories featuring women, we short-change women by not providing relevant education, and we also short-change men, by not providing examples of successful women.

The lack of visibility of successful women entrepreneurs is not only a deficiency in college education, but also in encouraging of women to gain proficiency in technologies, computer and telecommunications. Statistics show that more than 68 per cent of all users of the Internet are male and that most men use the Internet at work. Women need to have the same access and skill development opportunities as men, even though they may have interrupted careers, work from home, or fall into low-income ranges.

3. Ensure that all training programmes, seminars and materials feature the unique human and social capital endowments of women entrepreneurs and describe their resource acquisition process and organisational resource development.

4. Support and develop training centres providing software training, technology and telecommunications accessible to women entrepreneurs.

Recognition

While statistics show that women are an increasing economic force, their actions remain comparatively invisible in the media compared to their male counterparts (Baker *et al.*, 1997). Frequently print, radio and television media feature keys to success, or interview stories with successful male entrepreneurs, but they report ess often on successful women. Effectively, the contributions of women are undervalued, and consequently women are not taken as seriously as their male counterparts. This lack of visibility results in a lack of legitimacy for women as "successful' entrepreneurs (Baker *et al.*, 1997). Women then have to work harder to gain access to resources, convince resource suppliers (investors, loan officers, raw materials providers), government procurement officers and corporate customers that they are as competent as men (Brush, 1997). Even in the United States, women's share of government procurement continues to be small, less than 0.3 per cent of all contract dollars in 1983 and rising to only 10 per cent in 1988 (Lusgarten, 1994). "Invisibility" leads to women "not being taken seriously" with regard to their entrepreneurial endeavours (NFWBO, 1996; Brush, 1997). Public policy makers can lead the way by:

5. Featuring successful women entrepreneurs in marketing and informational brochures that promote government programmes or assistance.
6. Ensuring that government procurement opportunities for entrepreneurial businesses are equal for both male- and female-owned businesses.

Conclusion

This paper argues that women entrepreneurs come to venture creation with different endowments of human and social capital than their male counterparts. These differences are explained generally by the pervasive influence of similar social structures – work, family and organised social life – across all OECD countries. Social structures differentially affect women's human and social capital endowments, leading to unique combinations of organisational resources in their businesses. While many of the endowments that women entrepreneurs bring to their entrepreneurial endeavours can be assets in their new organisations, improved academic research, relevant educational materials and visibility of successful women entrepreneurs could significantly improve resource access and fill gaps. In particular, for the continued success of women's entrepreneurship into the 21st century, public policy needs to pay greater attention to research, relevant education and recognition.

Acknowledgements

The author is grateful to Dr. Patricia Greene and Dr. Myra Hart for their thoughtful comments and ideas in the development of this paper.

NOTES

1. The Babson Entrepreneurship Research Conference attracts the top 100-150 entrepreneurship researchers from all over the world. This annual conference requires an empirical paper as the criteria for admission, and is one of the few forums encouraging cutting edge empirical research. While the annual Proceedings are not comprehensive, they are representative of the current research focus in the field. This conference is held at an international location every third year.

BIBLIOGRAPHY

ADLER, N. and D. IZRAELI (eds.) (1994), *Competitive Frontiers: Women Managers in a Global Economy*, Blackwell Publishing, Cambridge, MA.

ADLER, N. (1994), "Competitive Frontiers: Women Managing Across Borders", in N. Adler and D. Izraeli (eds.), *Competitive Frontiers: Women Managers in a Global Economy*, Blackwell Publishing, Cambridge, MA, pp. 22-42.

ALDRICH, H. (1989), "Networking Among Women Entrepreneurs", in O. Hagan, C. Rivchun and D. Sexton (eds.), *Women-owned Businesses*, Praeger, New York, pp. 103-132.

ALLEN, S. and C. TRUMAN (eds.) (1993), *Women in Business: Perspectives on Women Entrepreneurs*, Routledge Press, London.

ANATAL, A.B. and KREBSBACH-GNATH (1994), "Women in Management in Germany: East, West and Reunited", in N. Adler and D. Izraeli (eds.), *Competitive Frontiers: Women Managers in a Global Economy*, Blackwell Publishing, Cambridge, MA, pp. 206-223.

ANDREWS, K. (1971), *The Concept of Corporate Strategy*, Richard D. Irwin, Homewood, IL.

BAKER, T., H. ALDRICH and N. LIOU (1997), "Invisible Entrepreneurs: The Neglect of Women Business Owners by Mass Media and Scholarly Journals in the United States", forthcoming in *Entrepreneurship and Regional Development*, Special Issue.

BARNEY, J. (1991), "Firm Resources and Sustained Competitive Advantage", *Journal of Management*, 17(1), pp. 99-120.

BECKER, G.S. (1964), *Human Capital*, Columbia University Press, New York.

BECKER, G.S. and G. GORDON (1966), "An Entrepreneurial Theory of Formal Organisations", Part 1: Patterns of Formal Organisations, *Administrative Sciences Quarterly*, December, pp. 315-344.

BOURDIEU, P. (1983), "Forms of Capital", in J. Richardson (ed.), *Handbook of Theory and Research for the Sociology of Education*, Greenwood Press, New York, pp. 241-258.

BRUSH, C.G. (1992), "Research on Women Business Owners: Past Trends, a New Perspective and Future Directions", *Entrepreneurship Theory and Practice*, 16(4), pp. 5-30.

BRUSH, C.G. (1997), "Women-owned Businesses: Obstacles and Opportunities", forthcoming in *Journal of Developmental Entrepreneurship.*

BRUSH, C.G. and B.J. BIRD (1996), "Leadership Vision of Women Entrepreneurs: An Exploratory Study", paper presented at the Babson Entrepreneurship Research Conference, April.

CHAGANTI, R. (1986), "Management in Women-owned Enterprises", *Journal of Small Business Management*, 24(4), pp. 18-29.

CHURCHILL, N. and J. HORNADAY (1987), "Current Trends in Entrepreneurial Research", in N. Churchill, J. Hornaday, B. Kirchoff, O. Krasner and K. Vesper (eds.), *Frontiers of Entrepreneurship Research*, pp. 1-22.

COLLIS, D. and C. MONTGOMERY (1995), "Competing on Resources: Strategy in the 1990's", *Harvard Business Review*, July-August, pp. 118-128.

COOPER, A. and GIMENO-GASCON (1992), "Entrepreneurs, Processes of Founding, and New Firm Performance", in D.L. Sexton and J.D. Kasarda (eds.), *State of the Art of Entrepreneurship*, PWS Kent, Boston.

EQUAL OPPORTUNITIES COMMISSION (1996), *Women and Enterprise: A Study of Self-employment in Northern Ireland*, Belfast.

GARTNER, W.B. (1985), "A Conceptual Framework for Describing the Phenomenon of New Venture Creation", *Academy of Management Review*, 14(4), pp. 696-706.

GLADE, W.P. (1967), "Approaches to a Theory of Entrepreneurial Formation", *Explorations in Entrepreneurial History,* 4(3), pp. 245-259.

GRANT, R.M. (1991), "The Resource-based Theory of Competitive Advantage: Implications for Strategy Formulation", *California Management Review*, Spring, pp. 112-135.

GREENE, P., C. BRUSH and M. HART (1997), "Resources in New Ventures: Dimensions and a Typology", Working Paper.

HAMMOND, V. and HOLTON (1994), "The Scenario for Women Managers in Britain in the 1990s", in N. Adler and D. Izraeli (eds.), *Competitive Frontiers: Women Managers in a Global Economy*, Blackwell Publishing, Cambridge, MA, pp. 224-242.

HANNINEN-SALMELIN, E. and T. PETAJANIEMI (1994) "Women Managers: Challenge to Management?", in N. Adler and D. Izraeli (eds.), *Competitive Frontiers: Women Managers in a Global Economy*, Blackwell Publishing, Cambridge, MA, pp. 175-189.

HOFER, C.W. and D.E. SCHENDEL (1978), *Strategy Formulation: Analysis and Concepts*, West Publishing, St. Paul, MN.

HOLMQUIST, C. (1996), "The Female Entrepreneur: Woman and/or Entrepreneur?", in NUTEK, *Aspects of Women's Entrepreneurship*, Swedish National Board for Industrial and Technical Development, Stockholm, pp. 87-114.

HOLMQUIST, C. and E. SUNDIN (1988), "Women as Entrepreneurs in Sweden: Conclusions from a Survey", in B. Kirchoff, W. Log, W.E. McMullan, K.A. Vesper. and W. Wetzel., Jr. (eds.), *Frontiers of Entrepreneurship Research*, April, pp. 643-653.

ITAMI, H. (1987), *Mobilising Invisible Assets*, Harvard University Press, Cambridge, MA.

INSTITUTE FOR SMALL BUSINESS (1996), *Myths and Realities: The Economic Power of Women-led Firms in Canada*, Bank of Montreal's Institute for Small Business, Montreal, Canada.

IZRAELI, D. (1994), "Outsiders in the Promised Land: Women Managers in Israel", in N. Adler and D. Izraeli (eds.), *Competitive Frontiers: Women Managers in a Global Economy*, Blackwell Publishing, Cambridge, MA, pp. 301-324.

IZRAELI, D. and N. ADLER (1994), "Competitive Frontiers: Women Managers in a Global Economy", in N. Adler and D. Izraeli (eds.), *Competitive Frontiers: Women Managers in a Global Economy*, Blackwell Publishing, Cambridge, MA, pp. 3-21.

JOHNSON, S. and D. STOREY (1993), "Male and Female Entrepreneurs and their Businesses: A Comparative Study", in S. Allen and C. Truman (eds.), *Women in Business. Perspectives on Women Entrepreneurs*, Routledge Press, London, pp. 70-85.

JOHANNISSON, B. (1996), "Existential Enterprise and Economic Endeavour: Women's Use of Personal Networks in the Entrepreneurial Career", in NUTEK, *Aspects of Women's Entrepreneurship*, Swedish National Board for Industrial and Technical Development, Stockholm, pp. 115-142.

KALLEBERG, A. and K.T. LEICHT (1991), "Gender and Organisational Performance: Determinants of Small Business Survival and Success", *Academy of Management Journal*, 34(1), pp. 136-161.

LERNER, M., C. BRUSH and R. HISRICH (1997), "Israeli Women Entrepreneurs: An Examination of Factors Affecting Performance, *Journal of Business Venturing*, 12(4), pp. 315-339.

LUSGARTEN, S. (1994), "Business Ownership as an Employment Opportunity for Women", Small Business Administration, 8035-OA-93.

NFWBO (1996), *Research Findings*, National Foundation for Women Business Owners, Silver Spring, MD.

OECD (1990), *Labour Force Statistics*, Paris.

OECD (1993), "The Rise of Women Entrepreneurs", *OECD Newsletter*, Paris.

PENROSE, E. (1959), *The Theory of Growth of the Firm*, John Wiley and Sons, New York.

PETERAF, M.A. (1993), "The Cornerstones of Competitive Advantage: A Resource-based View", *Strategic Management Journal*, 14(3), pp. 179-192.

POWELL, G.N. (1993), *Women and Men in Management*, Sage Publications, Newbury Park, CA.

PRAHALAD, C. and G. HAMEL (1990), "The Core Competence of the Corporation", *Harvard Business Review*, May-June, pp. 79-91.

RUMELT, R. (1987), "Theory, Strategy and Entrepreneurship", in D. Teece (ed.), *The Competitive Challenge*, Ballinger, Cambridge, MA, pp. 137-158.

SERDJENIAN, E. (1994), "Women Managers in France", in N. Adler and D. Izraeli (eds.), *Competitive Frontiers: Women Managers in a Global Economy*, Blackwell Publishing, Cambridge, MA, pp. 190-205.

STEINHOFF, P.G. and K. TANAKA (1994), "Women Managers in Japan", in N. Adler and D. Izraeli (eds.), *Competitive Frontiers: Women Managers in a Global Economy*, Blackwell Publishing, Cambridge, MA, pp. 79-100.

STEVENSON, H.H. and D.E. GUMPERT (1985), "The Heart of Entrepreneurship", *Harvard Business Review*, May-June, pp. 85-94.

SUNDIN, E. (1994), "Women's Entrepreneurship: A Reflection of Society at Large", in NUTEK, *Aspects of Women's Entrepreneurship*, Swedish National Board for Industrial and Technical Development, Stockholm, pp. 63-86.

STABER, Udo (1996), "Entrepreneurial Network Structures: Gender Differences and Performance Outcomes", Working Paper.

THE FLINDERS UNIVERSITY (1995), *Women in Small Business: A Review of Research*, report prepared for the Department of Industry, Science and Technology, The Flinders University School of Commerce, Australia.

THOMPSON, J. and J. HOOD (1991), "A Comparison of Social Performance in Female-owned and Male-owned Small Businesses", paper presented at the Annual Academy of Management Meetings, Entrepreneurship Division, Miami, Florida.

TOMER, J.F. (1987), *Organisational Capital*, Praeger Press, New York.

UNITED NATIONS (1991), *The World's Women 1970-1990: Trends and Statistics*, New York.

US GOVERNMENT (1995), *The State of Small Business*, US Government Printing Office, Washington, DC.

VOKINS, N. (1994). "The Minerva Matrix Women Entrepreneurs: Their Perception of their Management Style", in S. Allen and C. Truman (eds.), *Women in Business: Perspectives on Women Entrepreneurs*, Routledge Press, London, pp. 46-57.

WERNERFELT, B. (1984), "A Resource-based View of the Firm", *Strategic Management Journal*, 5, pp. 171-180.

WESTERBERG, L. (1996), "How Do Entrepreneurs in Female Sectors Express their Value?", in NUTEK, *Aspects of Women's Entrepreneurship*, Swedish National Board for Industrial and Technical Development, Stockholm, pp. 225-242.

Chapter 8

GLOBALISATION AND INTERNATIONAL TRADE

by
Andrina G. Lever
President, Women Entrepreneurs of Canada

Since the activity of women in SMEs and the economic impact of women-owned businesses in both developed and developing countries can no longer be doubted, much less ignored, women need to have greater involvement and representation. Women's current participation in the world economy is now only the thin edge of the wedge. Women do have different personal priorities and bring a new and different perspective to their businesses. Women need to be equitably represented at trade negotiations and in policy recommendations.

Globalisation has truly become one of the business and trade buzzwords of the late 20th century. The proliferation of bilateral and multilateral trade agreements and the conversion of many controlled economies to the free-market economy have been aided by such diverse business developments as global franchises, network marketing, product niches, Internet, technology and telecommunications, the movement of people and ethnic groups through migration and expatriates, multinational corporations, fashion, media and entertainment, standardization of legal applications and requirements, ISO 9000 and the adoption of English as the universal language of business.

Since it is already recognised that small and medium-sized enterprises (SMEs) account for more than 90 per cent of all businesses in the economies which make up the Asia Pacific Economic Co-operation (APEC) and 99 per cent of all businesses in the European Union, the effect of globalisation on these businesses cannot be underestimated. Fully 26 per cent of all OECD exports and 35 per cent of Asia's exports are produced by SMEs. Combine the above statistics with the fact that in all OECD countries, APEC economies, G7 countries, G-77 countries, NAFTA countries and within MERCOSUR and the European Union, the huge majority of women-owned businesses fall within the broad definition of SMEs and there is no doubt that globalisation and international trade will have a substantial effect on women-owned SMEs.

Women world-wide, even those in the most rural areas, will be affected by increasing globalisation and increasing global interdependence. As regions, rather than nations, are increasingly recognised as economic powers that transcend political jurisdiction, and as the newest technologies cross borders, businesses have to realise that they no longer live in isolated economies.

According to Dr. Rosabeth Moss Kanter (1995), we will be affected by the six great driving forces of globalisation:

- ◊ globalisation of industrialisation and technology;
- ◊ globalisation of finance, communication and information;
- ◊ globalisation of employment, work and migration;
- ◊ globalisation of human effects on biosphere pollution;
- ◊ globalisation of militarisation (nuclear and chemical weapons and their effects);
- ◊ globalisation of consumption, cultures and media.

It has to be recognised that the definition of an SME varies dramatically according to the various countries within the OECD and the sophistication of the economies in which they exist. Even within the G7 countries, the definition of an SME varies significantly, so for the purposes of this paper, the term SME is used somewhat generically.

The extent to which women-owned SMEs are globalised presently is limited. Indeed, the extent to which women-owned SMEs operate nationally within their own territorial boundaries is often very limited. A very high proportion of the businesses which women go into, or at least start in, are in the services sector, retail or in small businesses which serve the local market or community. Often these businesses do not lend themselves to international expansion. Also, traditionally, women-owned SMEs often do not reach a size where it is cost-effective to consider international business.

SMEs and globalisation

Companies winning in the global market-place share an emphasis on innovation, learning and collaboration. They:

- ◊ organise around customer logic;
- ◊ set high goals;
- ◊ select people who are broad creative thinkers;
- ◊ encourage enterprise;
- ◊ support constant learning;
- ◊ collaborate with partners (Moss Kanter, 1995).

However, the major challenge for globalisation and international trade is getting started! And, the challenges facing SMEs doing business in an international market can be overwhelming, even on a basic practical level:

- ◊ financial control;
- ◊ packaging of products so that they are more marketable;
- ◊ quality control ability;
- ◊ managerial capability;
- ◊ marketing and promotion.

It is also important to recognise the value of micro enterprises and home-based businesses even in the most developed economies. We cannot overlook that not all SMEs are in urban areas. Therefore, it is essential to develop initiatives that can empower and strengthen rural SMEs through activities and policies that focus not only on individual enterprises but also on groups of enterprises, co-operatives, associations and organisations.

The ability of SMEs to do business internationally is enhanced through access to world sourcing through technology – good suppliers and good ideas can be found anywhere in the world. Globalisation is possible through a variety of vehicles, such as:

- ◊ technology;
- ◊ partnerships;
- ◊ franchising;
- ◊ licensing;
- ◊ network marketing;
- ◊ foreign distributors;
- ◊ strategic alliances;
- ◊ joint ventures;
- ◊ supplying to other corporations.

The SMEs that globalise are those that specialise in products or services which lend themselves to foreign trade, the vehicles through which international trade can be carried out, or are suppliers to larger corporations that do business internationally. These factors can help to open doors and push women-owned SMEs to internationalise. Women-owned SMEs are also influenced by other factors such as:

◊ Lack of opportunity at home – the local market is small and saturated and, in order to continue to grow, the next step is to expand outside traditional boundaries.

◊ Increased movements of people and migration mean that a large part of the population have ethnic or cultural ties to another country where they may have personal contacts and already understand the culture and language, and recognise opportunities. This has happened a great deal with the opening up of the former eastern and central European countries, with many people returning to their countries of origin. In addition, many Asians who left Asia as part of the "brain drain" are now returning as part of the "brain gain".

◊ The increasing number of business networks and the maturity of existing networks such as *Les Femmes Chefs d'Entreprises Mondiales*, the International Federation of Women Entrepreneurs, the African Federation of Women Entrepreneurs, the International Alliance, the Women Senior Leaders Network, the CanAsian Businesswomen's Network, the Canadian Women's Business Initiative, Women in International Trade, women's divisions of Chambers of Commerce, professional or industry-specific organisations and a host of other organisations that may offer training, support, advice, access, networks, Internet access and business-matching opportunities.

◊ New emerging technologies are enabling women to cross borders – strong firms that develop and use technology are creating alliances and networks with numerous companies in many parts of the world.

◊ While the participation of women in traditional trade missions, such as the recent Team Canada missions undertaken in conjunction with Prime Minister Chretien, is still marginal, some countries are already promoting trade missions specifically designed for women – these trade missions are being carried out with the full support and participation of the government and may be industry-specific, focused on women doing business with women, or generic trade missions of women seeking business opportunities.

It is not accurate to say that women entrepreneurs are not risk takers – that is an oxymoron. Entrepreneurs are risk takers by nature. However, women tend to limit or mitigate their risk by doing a lot of research and asking a lot of questions. This characteristic which is inherent to women also applies to their strategies when they decide to do business internationally. The strategies employed to internationalise are no different from those used by men or traditional business. The difference is that women are very relationship-oriented and like to do business with people they know and with whom they feel comfortable.

Information

Finding information on foreign markets is no different for women than it is for men. SMEs interested in doing business internationally must take advantage of every avenue open to them, especially since financial resources may be fairly limited. Women rely heavily on relationships, referrals, networks and all the usual methods of identifying and investigating foreign opportunities.

The key to success in doing business internationally is to know the new market that the company is entering into, establish the right connections in that market and be able to be unique or competitive in that market. Since women-owned SMEs tend to be smaller than those owned by men, often they do not have the money or as many resources available to pursue international markets. In addition, until the business is large enough to be able to employ senior support staff, internationalisation requires the owner/operator of a company to travel. This can often be a challenge for women, given the responsibility that most women have with family obligations. Marketing may differ depending on the nature of the product or service that a woman-controlled business is providing, especially if it is geared to a woman's market.

Finance

The ability to finance exports of women-owned SMEs varies from country to country. In some countries it is more difficult for women to finance their companies even for domestic business purposes. In other countries the *perception* by women is that it is still more difficult for women to finance their companies, even when research and data dispute that perception. The challenges may result from the nature of the business of the SME or the size of the SME may prevent its taking advantage of certain government-insured or -sponsored programmes. In Canada, women do not get any special treatment for financing exports, and research tends to support the premise that they are not discriminated against. However, the perception of that reality is highly different.

One of the biggest challenges with respect to financing exports stems from the fact that most women-owned SMEs are in the service sector. Some banks are beginning to provide financing for foreign-based receivables. This is subject to SMEs being located in specific countries with sound economies, strong currencies and stable political systems. The fact is that they do not provide financing for foreign receivables in the service sector. Most countries have some sort of export development banks or corporations that will guarantee or insure foreign receivables but, again, these are not normally in the service sector and often are not available to women-owned SMEs due to the size or capitalisation of the exporting company.

For tax purposes it is advantageous for Canadian companies to keep their net value below a certain level. This may require some creative accounting which is totally legal. However, this will act against the company when producing financial statements and tax returns for the purposes of obtaining export development insurance or funds. The applying company must have a minimum net worth and show a steady record of growth.

International investments

International investments are usually complex and often risky. Different countries have different systems in place to protect investors and investments. Monitoring the use of the capital and the progress of the recipient company can be costly and difficult. There are different methods of assessment, accounting and accountability. It is always imperative to know the partners, investors or parties being invested in. The level of due diligence can vary dramatically.

Tax treatments at home and abroad are usually complex and require expert advice. Some countries have tax treaties which may allow the repatriation of profits without the penalty of incurring further taxes. Others may not. It is vital to know government policies with respect to foreign investment both in the country receiving the funds and in the country exporting the funds, *e.g.* the percentage of local ownership or participation required by law, the regulations for protecting the local population or indigenous peoples and possible restrictions on exporting funds.

Some countries have "immigrant funds" which allow foreign investment through a fund similar to a venture capital fund which then entitles the investor to specific privileges such as to live in that country or obtain a passport at an accelerated rate in exchange for investing a minimum amount of money in specific industries for a minimum period of time. Such plans have recently enabled Canada to attract huge amounts of foreign, particularly Asian, capital.

When investing abroad, political stability is an important consideration. Countries that have a history of political instability or inconsistent government policies present a much higher risk than those with a long democratic tradition. The recipient country's record of enforceability, official policies and confidence are critical to the advisability of investment. Tax treaties and potentially discriminatory practices should also be taken into consideration.

Foreign investors

For SMEs, the opportunity to open up their capital base to foreign investment is, quite simply, rarely an option. Attracting a foreign investor is competitive and is usually a result of personal contacts. Some countries only allow limited foreign investment; investment possibilities may be restricted to certain industries or it may be difficult to secure or monitor the investment. Often the amounts of money required by women are small and therefore not attractive to foreign investors. Investment would tend to come from family, friends, personal contacts and business relationships. Through organisations such as the OECD and other multilateral agreements, there is an opportunity to develop foreign investment funds for SMEs. Either incentives can be created to attract small but vital amounts of money as risk capital or small amounts can be placed in larger funds that will invest in SMEs but reduce administrative costs by consolidating administration. These concepts can be developed into a viable way of attracting and securing confidence in potential investments.

Access to international markets

Second- and third-tier exporting by supplying products or services to larger companies is a viable way of entering the international market while limiting risk. While the SME is dependent on the larger corporation to secure the international market, it can limit its risk and exposure by not having to bear the direct costs of developing and maintaining international contacts and markets.

International subcontracting and relocating can have an adverse effect on women-owned SMEs because these firms are often in the service sector, are small businesses with low overheads or are businesses that employ other women and low-cost labour. When other companies subcontract or relocate to economies with lower costs of labour, women are the first to be affected.

There is no reason why women-owned SMEs cannot be as competitive in a global market as any other SME. The greatest challenge is access to foreign markets and the cost of funds. Many successful women do not think of themselves as *women* in business but rather as business people. Women-owned SMEs are often competitive due to lower costs and expectations, but the cost of funding is often higher as they are charged more due to their lesser experience in negotiating.

Once in the international market-place, women have tremendous personal assets for doing business internationally. Most importantly, they are highly relationship-oriented; they take time to know who they are dealing with, are very intuitive and do not tend to do business with people they do not like. This combination works very well in cultures where it takes several meetings before business is even discussed. Women often possess excellent communication skills, tend to respect the people with whom they are dealing and are sensitive to cultural differences.

Contrary to traditional management styles which are determined by culture, such as the US business management style, the Japanese one or the German one, for example, women have their own management style which transcends geographical borders. This is a huge advantage which makes it easier to communicate and probably stems from the common background of women and how they are conditioned. Regardless of culture, women are groomed to be the care givers, nurturers, mentors for children, mothers and wives, and to balance the many roles we have to play.

As business women, we face the same challenges and responsibilities that women face world-wide: balancing our lives and not having wives to do all those things which business men have wives to do! The degree to which women face those challenges may vary from country to country depending on culture, economics, race, religion, politics, ethnic backgrounds, but the fact remains that women are highly intuitive and have much in common with other women simply because they are women. Women tend to be less ego-oriented than men and are often able to transcend cultural difference because of this. This common background is a huge ice breaker when women are together. The power of these skills should not be underestimated.

Often women do not have the experience, skills or confidence to be as aggressive as may be necessary to "close the deal". Increased training and exposure can prepare them for international trade and globalised activity. However, women can also influence the way business is done internationally. We manage our companies differently – more like we manage our families. Women are more environmentally and socially aware and responsible. We look at the effect of business on communities. We are concerned about the future and its effect on the lives of our children. And, contrary to the popular opinion of men – and financial institutions will confirm this – we are more fiscally responsible – we borrow less money and we pay back our loans.

We are also aware that we need to keep in mind a vision of the future that reflects the needs and aspirations of young people. We know we will be judged on the condition in which we leave the world. We think about how our decisions affect the children and young people in our lives.

However, in order to be successful internationally it is still necessary to have access to, and knowledge of, international markets. It is necessary to make sure that the products manufactured are appropriate for the international market. Women are major players in the area of style and fashion, but they must know what is saleable in another country. I personally have visited a knitting factory in Vietnam with excellent export potential and which is already exporting in a small way to Japan and Australia. Pricing is a big advantage but, in order to compete successfully, the factory must have access to international design preferences, must manufacture sweaters suitable for their respective foreign markets and must be consistent in quality. In many countries, especially those of the South, the handiwork is beautiful but the colours or designs are too ethnic for tastes in the North. With design assistance and quality assurance, these products can be adapted to compete in the global market-place.

Service SMEs

Since such a large number of SMEs are in the service sector and possess their own set of challenges, it is worth mentioning that their trade promotion is of particular importance. Dr. Dorothy Riddle, President and CEO of Service-Growth Consultants, Inc., outlines the following advice to service companies (Riddle, 1996):

◊ trade in services is growing rapidly, especially trade in "business services", with the United States, Japan and Hong Kong the lead APEC markets;

◊ service SMEs are competing with some very large service transnationals (which comprise 49 out of the top 100 corporations world-wide), dominated by Japanese service conglomerates;

◊ in order to compete successfully, service SMEs have to be niche players, able to develop new services and respond rapidly to new market opportunities;

◊ to export successfully, services SMEs have to promote first the competitiveness of their capabilities and then the competitiveness of their services – for professional service SMEs in particular, marketing must be done by senior professionals, who are not necessarily good marketers;

◊ promotion of capabilities means creating credibility or profile with potential new clients through objective verification of capabilities, providing personal guarantees and managing one's presentation image;

◊ successful profile-building activities include joining local trade or service industry associations, speaking at major trade events, developing media coverage and subcontracting to well-known firms;

◊ the operational strengths of a particular service SME will determine the type of market in which the firm is likely to do well;

◊ the most efficient way for service SMEs to enter a new market is in conjunction with a local partner, so identifying potential partners and developing a partnering relationship are key strategies.

Technology

Deregulation and multimedia technology is giving birth to new business and niche markets, and technology is becoming the great equaliser for women in business. We have a journalist in Canada that believes that men and women were truly equal until the discovery of the axe which required brute strength to swing. Since then, men have dominated. However, he proclaims that the advent of technology will now level the playing field once again!

The potential for technology in international trade is enormous. However, in both developed and developing countries, the infrastructure to provide access to technology has to be in place and must be available at a reasonable cost. This relies heavily on telecommunications and government policy.

Technology itself is gender-blind. The issue is to get more women to be comfortable with technology and learn to use it as an integral part of their business. At the technological level, opportunities exist to:

◊ provide training;

◊ provide distance learning;

◊ simplify procedures;

◊ eliminate cumbersome paperwork;

◊ make information available, accessible and inexpensive;

◊ reduce time demands;

◊ assist in small-business training programmes;

◊ create a global network of information for women;

◊ create a global market-place for services and products;

◊ develop the use of electronic conferencing.

The extent that women use technology in doing business in general is relatively limited but certainly on the increase. There is much to be done and huge opportunities exist. It is not necessary that women understand the technology itself, but rather that they recognise the need for it and hire the talent that can deliver the technology they require. In some countries the cost of technology and access to the Internet is limited or expensive given the high per unit costs of telephone services. Possible unreliable power supplies can dramatically affect the availability and efficiency of technology.

According to the Harvard School of Management (1994):

"We are standing on the verge, and for some it will be the precipice, of a revolution as profound as that which gave birth to modern industry. It will be the environmental revolution, the generic revolution, the materials revolution, the digital revolution, and, most of all, the information revolution. Entire new industries will soon be born. Existing industries will be profoundly transformed and new opportunities will emerge. Many of these mega-opportunities represent billions of dollars in potential future revenues. Many will fundamentally transform the way we live and work. Each of these opportunities is also inherently global. No single nation or region is likely to control all the technologies and skills required to turn these opportunities into reality. Markets will emerge at different speeds around the world and any firm hoping to establish a leadership role will have to collaborate with and learn from leading-edge customers, technology providers and suppliers wherever they are located.

The future is now. The short term and the long term are not clearly delineated anymore. They are tightly intertwined. Although many of tomorrow's mega-opportunities are still in their infancy, companies around the world are at this moment competing for them. Alliances are being formed, competencies are being assembled and experiments are being conducted in nascent markets – all in hopes of capturing the world's future opportunities."

So, what does this mean for women? It means a change in the way we do business. It means more challenges – to act swiftly and to partner with others. But women are used to rising to challenges!

International perspectives

Since the issues that are being presented to the OECD during this Forum are so similar in nature to those that have been presented to other multilateral organisations, it is worth including the following information.

APEC (Asia Pacific Economic Co-operation)

In September 1997, Canada will be hosting the Asia Pacific Economic Co-operation SME Ministerial meeting in Ottawa. For the two days prior to that meeting, women will gather from all of the 18 APEC economies to focus on the general SME themes of APEC:

- ◊ access to finance;
- ◊ technology;
- ◊ business skills development market access;
- ◊ access to information.

These themes are identical to those identified for this OECD Summit and for other similar fora. Having so consistently identified these themes, it is now necessary to address them and for governments to realise that they have a responsibility to help create the right conditions for equitable economic prosperity and to provide the proper framework for trade and investment liberalisation measures to succeed.

In her paper to the APEC Women Senior Leaders Network in Manila in October 1996, Dr. Lorna Marsden, President and Vice Chancellor, Wilfrid Laurier University, Canada, stated (Marsden, 1996):

"The arrival of the World Trade Organisation in 1995 signalled the end of the beginning of globalisation – through GATT, the WTO and other international trading treaties and multilateral agreements, national governments have inserted themselves or have been inserted into trading relationships in order to facilitate private sector trades and boost national economies. While concerns are being raised about national sovereignty as the WTO rules extend into areas of domestic policy, no government can now afford to be outside these arrangements.

Movements have an obligation to all their citizens – they need to inform all interested parties of trading opportunities, to publish records of agreements and their process and to manage their trade involvements through the public service. This includes women. One of the fields of investigation in this area is the forms and processes of national governments which permit the greater open involvement of women in advisory, elected, expert or management roles.

The presence of women ... and the rising importance of trade in all industries opens new opportunities to concern ourselves with the impact of trade on the economic lives of women in the national economy at all levels and on the policies that shape these relationships. It is not enough to say that policy takes women into account. In order to provide a good foundation for estimating the impact of trade and other relationships, it is essential that government statistical agencies examine the quality and coverage of the data they collect and the standardized measures to ensure that we can analyse the extent to which female-intensive industries are changed by the new trading relationships. In addition to industry and sectoral groups as they are presently defined, these need to include studies of the informal sector, home workers and any other significant female-dominant segments of the economies.

- ◊ governments can be instrumental in creating or supporting sectoral organisations and firms that are inclusive of female-intensive parts of the economy as well as supportive of women entrepreneurs;
- ◊ governments can also provide special opportunities to women in trade: training courses, trade fair opportunities, trade mission experience, mentoring by experienced trade officers and access to expert advisors;
- ◊ creating consultative opportunities to governments and their departments while policy is being developed is an important contribution of governments, as is supporting the appointment of women to senior positions in the public and private sectors;
- ◊ governments play a crucial role in creating pairing and partnerships in particular sectors which can make a difference between competitive success and failure."

Susan Joekes and Ann Weston wrote in their monograph for UNIFEM (Joekes and Weston, 1994) that the questions are posed:

- ◊ Are the new trade rules likely to have a particular impact on women or on particular groups of women?
- ◊ Are there policies in place to deal with these gender impacts?
- ◊ What types of rules might help to promote women's entitlements and capabilities?
- ◊ Are there mechanisms to ensure that the gender dimensions of trade are addressed?

Joekes and Weston also identify the issues facing trade policy makers, such as the traditional issues of tariffs, non-tariff barriers and commodity prices, and the new issues of services, intellectual property rights and foreign investment rules. Rising rapidly on the agenda are the environment, labour standards and competition policy. They examine the contentious issues of the struggles between legislative and executive powers and between levels of government; the transparency of policy making in governments and the participation of agencies representing women's interests; the presence of independent review bodies; and the role of the private sector. This report is a very useful guide to the general issues in international trade and their potential impacts on the lives of women; it concludes by citing seven ways in which advocates can promote gender issues in trade:

- ◊ monitoring employment conditions;
- ◊ using trade agreements to promote gender equity;
- ◊ seeking improvements in the informal sector;
- ◊ improving access to credit;
- ◊ training and marketing support services;
- ◊ compensating women displaced by trade;
- ◊ helping women entrepreneurs reorient their output to export opportunities;
- ◊ improving education and training facilities for women.

Two other issues stand out as being of great importance to any work in international trade. The first is the relation of women to the new transforming technologies. These are the robotic, materials, computer and biotechnology inventions which are pervasive in the workplace and in the consumer market-place. In the financial sector, these new technologies are already affecting the lives of millions of workers. But, in particular in transportation, communication and automated production, the sweep of technological change through industry has changed the nature of competition among firms and countries and has certainly changed the skill levels of workers.

In addition, there is a need to retain the knowledge of craft and skilled work that women have so often carried out – in herbal and other forms of medicines; in sewing, weaving and other traditional and decorative skills; in cooking and domestic skills. Technology, world mandates and corporate dominance are causing these skills to be in danger of being lost in some places. The trade aspects of intellectual property rights have already been identified at a high level and will bear fruit. Eventually a very high premium will be put on many of these skills, as "hand made" goods and "traditional" services become the high-priced privileges of the well-to-do. A formal agreement to retain, with government support, the traditional knowledge mostly but not exclusively practised by women would be a useful outcome of this Forum.

Partnerships among women-led firms participating in these various fora signed at the fora would make real to governments the power of women in these industries, raise the profile of women in the public perception and advance the general cause of trade."

Dr. Marsden concludes by stating that it is important:

- ◊ to get the right issues on the agenda;
- ◊ to get the right people and more women in the delegations;
- ◊ to get an action plan with resources committed to the achievement of the goals set.

I quote so heavily from Dr. Marsden's paper because I endorse her comments and submit that the same principles apply to us now at the OECD.

European Union

According to Francisca Nennasar Tores (1995), author of an opinion for the Committee on Women's Rights in the European Commission's Annual Report of 1995, women constitute more than half of the world population, are the main producers of wealth, carry out two-thirds of all work performed but only receive one-tenth of the world's income and own 1 per cent of property. In Europe, 80 per cent of the poor are women.

Although the European Union did not recognise that one in three new SMEs in Europe is being created and managed by women, in a report presented by the European Commission for the Madrid European Council in 1995 (European Commission, 1995), it was acknowledged that:

> "There is now widespread recognition within the EU that small and medium enterprises are the key factor for generating employment opportunities and growth in the European Union.
>
> SMEs account for 99.8 per cent of all companies, 66 per cent of total employment and 65 per cent of business turnover in the European Union. In particular, those at the smaller end now play a key role in terms of growth and generate an above-average share of new jobs.
>
> European SMEs' full potential for growth and employment creation is not being properly realised. On average 50 per cent of SMEs in the EU fail within the first five years of start-up... In addition, 30 000 enterprises and 300 000 jobs are threatened every year simply because SMEs fail to overcome the difficulties in the transfer of ownership.
>
> Among the factors offering new opportunities for SMEs are the growing importance of the service sector (in particular business services), the increasing role of technologies favouring small-scale production, easier access to global markets (including the Information Society) and the move towards outsourcing, subcontracting and downsizing by large firms. SME development is also essential for maintaining social and economic cohesion and assisting rural development."

Some of the major barriers to stronger development of SMEs include:

- ◊ excessive costs arising from regulations such as VAT or GST compliance and inheritance taxes;
- ◊ limited creation of new-technology-based firms;
- ◊ the cost of labour and lack of training;
- ◊ under-capitalisation due often to the difficulty of obtaining loans;
- ◊ barriers to internal markets;
- ◊ lack of knowledge about global market opportunities.

The European Union has already identified a series of policies and concrete actions which are essential to strengthen SME job creation and growth potential:

- ◊ think small first – reduce the structured bias against SME creation and development resulting from excessive bureaucracy;
- ◊ facilitate the successful transfer of businesses;
- ◊ ensure better SME representation in policy making;
- ◊ encourage the creation of new-technology firms;
- ◊ facilitate innovation and research by technology-intensive SMEs;

- ◊ promote the use of information technology by SMEs;
- ◊ ensure lower non-wage labour costs;
- ◊ encourage an entrepreneurial culture;
- ◊ adapt training to the needs of SMEs and encourage good management practice;
- ◊ introduce incentives for SMEs to invest in training;
- ◊ reduce tax discrimination between debt and equity financing for SMEs;
- ◊ improve access to credit;
- ◊ facilitate access to credit through the European Investment Fund;
- ◊ facilitate the creation and operation of European capital markets for SMEs;
- ◊ remove entry barriers to markets and reduce distortions resulting from state aid;
- ◊ improve access to the internal market;
- ◊ help SMEs to internationalise.

The six priority actions targeted for Member States and the EU are as follows:

- ◊ reduce red tape which hampers European entrepreneurship;
- ◊ ensure better involvement of SME organisations in the decision-making process;
- ◊ help finance SMEs which create new jobs;
- ◊ reduce market distortions and inefficiencies;
- ◊ vigorous action to promote research, innovation and training to assist SMEs;
- ◊ enhance competitiveness and internationalisation of SMEs.

SMEs are so vital to the economy that we cannot overlook them – there are 18 million in Europe alone and one-third are owned and operated by women! That represents six million SMEs, and the number is on the rise. There is an urgent need to eliminate red tape and heavy paperwork in order to free up time, people and resources to concentrate on business.

Canada

An initiative has been started by the Canadian government called the Canadian Women's Business Initiative, aimed specifically at women to encourage them to expand their businesses initially to the mid-Atlantic US states, but ultimately farther. Historically, trade officers have focused on different industry sectors in order to encourage international trade. Only recently as a direct result of the OECD Regional Preparatory meeting held in Washington, DC, at the Canadian Embassy, was the Canadian Women's Business Initiative developed whereby the Canadian government is now focusing on women-owned businesses as a sector to promote.

The Canadian Government is committed to implementing gender equality at all levels of national, international and multilateral activities. As stated in the Federal Plan for Gender Equality (Status of Women Canada, 1995), the federal government will continue to focus attention on women's issues in:

◊ pursuing gender equality objectives in various multilateral fora;

◊ promoting the reform of international and multilateral institutions and mechanisms in order that their policies and programmes better reflect and meet the needs of women;

◊ promoting the integration of gender equality issues in activities of international organisations, including integrating gender analysis in all work with the OECD.

Strengthening the full participation of women as equal partners in sustainable development: Women in Development (WID) is one of the six priorities of Canada's Official Development Assistance programme administered by the Canadian Development Agency. The goal of CIDA's WID and gender equity policy is to strengthen the full participation of women as equal partners in the sustainable development of their societies. The policy has two distinct but interrelated components:

◊ the effective integration of gender equity considerations into all development initiatives at the policy programme and project levels;

◊ ensuring that women have the opportunity to participate in and benefit from all initiatives as equal and active partners.

Its objectives over the next five years are:

◊ to encourage, respond to and support initiatives within and among the developing countries to:

- increase women's participation in economic, political, social and environmental decision-making processes;
- improve women's income levels and economic conditions;
- improve women's access to basic health and family-planning services;
- improve women's levels of educational achievement and skills;
- protect and promote the human rights of women.

◊ to promote the elimination of discriminatory barriers against women.

With such a noble mandate already committed to by the Canadian federal government, each of these activities include by assumption women-owned SMEs and their participation in globalisation and trade.

Trade and gender declarations

The timing of this OECD Forum is perfect. Since June 1996, at least four major declarations have been drafted with respect to gender and trade liberalisation:

◊ 26 June-3 July 1996 – Accra, Ghana;

◊ 2-4 October 1996 – Manila, Philippines (presented to President Ramos of the Philippines for presentation to APEC leaders);

◊ 14-16 November 1996 – Bangkok, Thailand;

◊ 9 January 1997 – Miami, Florida (presented to G-77 leaders in San Jose, Cost Rica).

Policy recommendations

Finally, when reviewing policy recommendations for organisations such as the OECD to take into account, the following questions must be considered and discussed in order to offer concrete recommendations:

◊ How can science and technology assist SMEs?

◊ The gender dimension in policies for the development of SMEs.

◊ The need for policies that protect against unfair business practices but allow the development of sound competition.

◊ Access to markets.

◊ Trade liberalisation and facilitation policies which affect women-owned business especially micro business and low-cost, unskilled labour.

◊ Trade pacts negotiated by men with few women at the table – women's perspective is different and valuable.

◊ Women need knowledge and training.

◊ Financing of foreign service receivables, especially since so many woman are in the service sector.

◊ Simplification and standardization of customs procedures and certificates of origin.

◊ Immigration, travel permits and visas allowing easy and efficient processing for business travellers.

◊ Eligibility for export development insurance and finance.

Governments need to actively seek and act on the advice of the private sector, and in particular SMEs, in setting policy directions. Since the activity of women in SMEs and the economic impact of women-owned businesses in both developed and developing countries can no longer be doubted, much less ignored, women need to have greater involvement and representation. Women's current participation in the world economy is now only the thin edge of the wedge. Women do have different personal priorities and bring a new and different perspective to their businesses. Women need to be equitably represented at trade negotiations and in policy recommendations.

BIBLIOGRAPHY

EUROPEAN COMMISSION (1995), Madrid Council.

HARVARD SCHOOL OF BUSINESS MANAGEMENT (1994), report.

MOSS KANTER, Dr. Rosabeth (1995), *World Class*, Simon & Schuster.

RIDDLE, Dr. Dorothy (1996), President and CEO, Service-Growth Consultants Inc., speech delivered in Manila, September.

JOEKES, Susan and Ann WESTON (1994), *Gender and the New Trade Agenda,* Women, Ink., UN Development Fund for Women, New York.

MARSDEN, Dr. Lorna (1996), paper presented to Women Senior Leaders' Network launch, Manila, October.

TORES, Francisca Nennasar (1995), Committee of Women's Rights in the European Commission, Annual Report.

Chapter 9

WOMEN STARTING NEW BUSINESSES: THE EXPERIENCE IN NORWAY AND THE UNITED STATES

by
Nancy M. Carter
Morrison Chair in Venture Creation, and Professor of Management
Marquette University, United States
and
Lars Kolvereid
Bodo Graduate School of Business, Norway

Comparisons between women and men engaging in entrepreneurial activities suggest that they share many of the problems encountered in establishing new ventures. There are, however, some areas where experiences differ. In particular, financing may be a more significant issue for women. Women appear to have fewer financial resources available at start-up. They tend to come from households with lower income levels and they register intentions to start the firm with considerably less financial support, particularly debt. These findings may suggest that women are not as aware of the magnitude of resources required to establish a business and/or that they have less access to financial resources.

The findings among US nascent entrepreneurs suggest that the entrepreneurial climate may play an important role in triggering women's decisions to select self-employment. Women who became discouraged during the early stages of developing the business and gave up were much more likely to view the climate as negative. To the extent that perceptions of climate can "trigger" the entrepreneurial process, programmes should be developed that fully represent the opportunities available to women. Profiles of women business owners must be more readily available and widely disseminated to facilitate role modelling behaviour.

Women-owned businesses represent one of the fastest growing segments of economies around the world. A report of the House Small Business Committee in the United States, for example, demonstrates that between 1970 and 1988 the number of women-owned businesses grew over six-fold from less than 5 per cent of all US businesses in 1970 to over 30 per cent by 1988 (US House of Representatives). This rate of growth is especially remarkable since women-owned businesses are likely to be underestimated in many government censuses. Preliminary data from a 1997 national panel study of business start-ups in the United States reveals that in 7 per cent, or more than 6 million households, one or more persons is involved in a new start-up. Of these, one in three entrepreneurs starting businesses are women.

Despite this growth rate, many argue that significant barriers still remain for women in establishing and growing businesses. These barriers may explain why businesses owned by women tend to be smaller, with receipts substantially lower than those of men-owned businesses in comparable industries.

Sources of data

Data from Norway and the United States were examined in this paper to isolate the particular challenges women face in starting and managing their businesses during the critical early years. Nascent entrepreneurs, or persons taking steps to start new businesses but who have not yet succeeded in making the transition to new business owners, were identified in both countries. In the United States, data on nascent entrepreneurs came from two representative samples; one of 683 adult residents in the state of Wisconsin (hereafter referred to as WI-Nascents), and one of 1 016 adult residents of the United States.[1] The procedure started with a random selection of households, followed by a random selection of an adult member of the household. Data were collected through a telephone interview. Respondents were asked: "Are you alone, or with others, now trying to start a business?" Those responding "yes", and who reported initiating steps to get the business running, were retained in the sample. Follow-up interviews were completed with respondents from both samples approximately 18 months after the first data collection.

In Norway, data on nascent entrepreneurs came from a random sample of 9 533 adults contacted by a professional survey company. In the phone interview 203 responded "yes" to the question about whether they were trying to start a new business (hereafter referred to as Norway-Nascents).

Women's involvement in the entrepreneurial process

Findings from the US survey support research demonstrating that men starting businesses outnumber women almost two to one. Sixty-four per cent of the US nascent entrepreneurs were male and 36 per cent female. The gender disparity of women in the entrepreneurial process is particularly notable when one considers that, in 1990, 57 per cent of all American women age 16 and over were in the labour force. The disparity is even more pronounced among respondents in the Norwegian survey. Seventy-six per cent of the respondents in Norway were male, and only 24 per cent female.

Factors that influence the creation of new firms

Conditions that have been viewed as stimulating the birth of new businesses range from labour market conditions, industrial competition, literacy and technological innovation, to personality characteristics of individuals. For entrepreneurs in the early stages of establishing their businesses, two sets of conditions seem especially critical; incubator influences – factors that encourage the "propensity to act"; and facilitative factors – conditions that foster the gestation process once start-up has been initiated.

Incubator influences

Incubator influences are factors that stimulate the "triggering" event or the propensity to select self-employment as a viable work alternative. Four influences are thought to be particularly instrumental: *i)* perceptions regarding favourableness of the environmental context in supporting entrepreneurs; *ii)* life-stage position of potential entrepreneurs; *iii)* employment status; and *iv)* work values or career intentions.

Entrepreneurial climate

The extent to which potential entrepreneurs view the environmental context surrounding them as one supportive of entrepreneurial efforts may influence whether or not they initiate start-up efforts. In the Wisconsin study, the nascent entrepreneurs were asked questions about how favourable they thought the local environmental context was for starting new businesses. For each question respondents had four choices: Strongly agree, Agree, Disagree and Strongly disagree. They were given values from 1 to 4 so that the higher the number, the more positive the response about the state as a place to start a business. These data were used to constructed a "climate" assessment.

Overall, women were more positive in their judgements than men. Significant differences were found in three aspects of the environment; perceptions regarding government assistance, perceived level of social support and favourableness of obtaining financial help. Table 1 displays the findings. Women nascent entrepreneurs viewed the assistance provided to men by state and local governments significantly more positively than men viewed that support (mean values = 2.76 for women *vs.* 2.38 for men). Women also felt more strongly that people would start new businesses if it were not so complicated to obtain help and approval from the government (mean values = 3.33 for women *vs.* 2.98 for men).

Among the nascent entrepreneurs interviewed, some had tried to get a business started, but had become discouraged and given up. Among this discouraged group, women indicated feeling the least social support for their business initiatives. Discouraged women were significantly more likely to think that if a person's business fails, they would never be respected again (mean values = 2.40 discouraged women *vs.* 1.96 discouraged men entrepreneurs, or 1.93 women nascent entrepreneurs. They also were less likely than other entrepreneurs to think, having tried and discontinued the process, that they would repeat the entrepreneurial initiative, and felt that more women would start new firms if they could obtain some financial help (mean values = 3.26 for discouraged women *vs.* 3.04 for discouraged men).

Table 1. Nascent entrepreneurs

		Norway		United States	
		Men	Women	Men	Women
GENDER		76%	24%	64%	36%
INCUBATOR INFLUENCE					
Entrepreneurial climate[1]	Government provides support for men			2.38	2.76
	More starts if not so complicated			2.98	3.33
	Lose respect if business fails[2]			1.96	2.40
	More starts with financial help[2]			3.04	3.26
Age				37 yrs	38 yrs
Marital status[1]	Married/living with partner			64%	71%
Children living at home[1]	Under 18 years old			65%	60 %
Labour force status	Full time job	81%	38%	68%	41%
	Part time job	10%	31%	7%	24%
	Self-employed			14%	24%
	Unemployed	7%	29%	7%	0%
	Retired			0%	6%
	Student/Disabled/Other			4%	6%
	Varies	2%	2%		
	Current owner manager	42%	29%		
Work values[1]	Autonomy/impact			3.4	3.4
	Task interest			3.3	3.4
	Wealth			2.7	3.0
	Community interest			3.0	3.0
Career choice	Self-employed	69%	71%		
	Organisationally employed	21%	24%		
	Don't know	10%	4%		
Sales expectations (median)[1]	1st year			$30 000	$20 000
	5th year			$55 000	$35 000
	10th year			$200 000	$60 000
FACILITATIVE FACTORS					
Human capital					
Work tenure				14 yrs	11 yrs
Start-up experience				21%	21%
Discouraged entrepreneur		31%	13%		
Financial capital[1]					
Expected sources	Entrepreneur's money			33%	57%
	Informal money			12%	8%
	Formal money			47%	26%
	Team money			8%	9%
Household income	Up to $10 000/yr	2%	0%	4%	6%
	$10 000-29 000/yr	8%	27%	23%	39%
	$30 000-49 000/yr	22%	22%	42%	33%
	$50 000-79 000/yr	46%	27%	23%	17%
	$80 000-up/yr	18%	9%	8%	5%

1. Wisconsin nascents only.
2. Discouraged nascent entrepreneurs.

Overall, women were more positive than men in their perception that the environmental context supports entrepreneurship. But for those women who had given up trying to establish their businesses, the climate was perceived as negative. These women associated the greatest stigma to failure, indicated that they were least likely to try another entrepreneurial initiative and were most dissatisfied with access to financial resources that would be helpful in starting their businesses.

Life stage

The life stage of entrepreneurs also may influence the decision to initiate start-up activities. Life stage includes the entrepreneur's age when committing to the new business, marital status of the entrepreneur and the number of children living at home. Of interest is whether the life stage of women establishing entrepreneurial ventures differs from that of men.

Data from the WI-Nascent study indicate that there is little difference in the ages of men and women nascent entrepreneurs. Both groups tend to be about 35-40 years old. There is no significant difference in the marital status of the entrepreneurs. More than 50 per cent of both groups are married. Similarly, there is no significant difference in the number of children living at home. Whereas previous research has suggested that life-cycle stage affects the growth of established new ventures, the WI-Nascent data reveal few differences in the life stage of women and men as they embark on initiating new businesses.

Employment status

Discussions regarding whether people are "pushed" into starting new businesses because of unemployment or "pulled" by opportunities in the environment, raise the question of whether the stimulus differs for men and women. Data for US entrepreneurs reveals little support for the supposition that unemployment leads to creating new firms. Only 7 per cent of men reported they were unemployed and looking for work prior to attempting to start the current business, and none of the women nascent entrepreneurs. In contrast, almost 30 per cent of Norwegian women reported being unemployed before starting the current new venture. Similarly, there are significant gender differences among the work patterns of the entrepreneurs, with women nascent entrepreneurs in both countries less likely to be working full-time. Almost 25 per cent of US women reported working part-time prior to the start-up initiative and more than 30 per cent of Norwegian women. Women's higher rate of part-time employment may indicate that these women are underemployed relative to men and, having fewer work alternatives, may turn to self-employment. In the United States 24 per cent of the women report being self-employed prior to starting the current new venture, while 29 per cent of Norwegian women report being current owner-managers (either as their part-time or full-time job).

Work values

Previous studies have suggested that the reasons for selecting self-employment or entrepreneurship as a career alternative may differ by gender. Men have been seen as establishing new businesses out of a desire to be entrepreneurs, or to not work for someone else. Women, on the other hand, have been viewed as wanting to create new ventures where they can balance work and family (Brush, 1992). Respondents to the Wisconsin survey were asked a series of questions about their work interests or values. A factor analysis of these items revealed four distinct dimensions: autonomy or independence; task interest; wealth or income; and a desire to stay in the community. A comparison of the work values by gender provided mixed support for previous research findings. As expected, the nascent men entrepreneurs ranked wealth and prestige highest (mean value = 2.7, where a low value corresponds to a more positive orientation), but the mean value was not statistically different from that of women. Similarly, there were no significant differences between men's and women's desire for autonomy and independence or for pursuing challenging, interesting work.

In the Norwegian survey respondents were asked about their career preferences. The findings reveal that women (71 per cent) were just as likely as men (69 per cent) to want to be self-employed rather than work for an established organisation (women = 24 per cent *vs.* men = 21 per cent). This is in stark contrast to the prevalence rate reported above indicating that more than three times as many men than women in Norway are actually starting their own businesses.

Overall the findings fail to support previous contentions that women have different work values than men. Both groups appear to view self-employment as a desirable career option and as a way to provide wealth or build an estate for their families. And they want to do this while remaining in the area where they presently reside. Presumably the relationships they have established in the local environment are as important to them as wealth creation. Despite little difference in work values, there is a distinct difference in the number of women and men starting new businesses.

Intentions/aspirations

The personal aspirations and intentions of entrepreneurs also are thought to play a key role in predicting the process of establishing new businesses and their subsequent success. The role of work values as a stimulus to selecting self-employment was evaluated above. Contrary to previous research findings, no significant differences were found in the values women and men placed on achieving wealth and prestige as an outcome of their work. But what does "wealth" mean, and does the meaning vary by gender? In contrast to a stimulus for initiation (a measure useful for explaining why there are fewer businesses owned by women than men), differences in the perception of what constitutes wealth may explain why the size and scope of women-owned businesses are substantially smaller than those owned by men.

The US surveys provide data on the entrepreneurs' expectations for both the size and growth plans of their new businesses. Specifically, the respondents were asked: "If the business is launched, what sales would you expect at the end of the first year? At the end of the fifth year? At the end of the tenth year?" Comparison of the median expected sales reveals only slight differences in what the entrepreneurs anticipate achieving by the end of the first year. Men reported they expected to have approximately $30 000 in sales in comparison to women's expectation of $20 000. By the end of the fifth year of operation, men expected average sales to reach $55 000 in comparison to women's expectation of $35 000. By the end of the tenth year, men expected $200 000 in yearly sales in comparison to women's expectation of $60 000. In other words, by the tenth year of operation, men expect sales almost three times higher than those expected by women. Clearly, the vision held by men and women about the future size of their businesses differs. As vision represents intentions, the size differential in women- and men-owned businesses seems to be "set at the start" as argued by Birley (1986). Women appear to anticipate having smaller businesses over time than do men.

Facilitative factors

Once potential entrepreneurs have made the decision to initiate the start-up process, they enter into the gestation stage of the firm's life-cycle. Studies on how pre-organisations or organisations-in-creation become fledgling new businesses have sought to identify factors that facilitate the transition, as well as systematic barriers or obstacles that may hinder the emergence of new ventures. Some have speculated that certain sub-populations of emerging ventures may be especially vulnerable during this early development stage. Women may represent one such group. Research on established women-owned businesses suggests three reasons why nascent women entrepreneurs may experience early difficulties. Women have been found to have: *i)* fewer human capital resources to bring to the firm creation (Bates, 1985); *ii)* fewer financial resources (Cromie and Birley, 1991); and *iii)* women may not utilise assistance programmes offered to entrepreneurs by government or private sources as readily as do men.

Human-capital resources

Human capital derives from the investments individuals make in themselves, often through education, training and skills learned through work experience. Women have been viewed as having lower levels of human capital as a result of cultural stereotyping. Human capital which comes in especially useful in starting a new business may include how much experience individuals have had in working for others. Also relevant is whether they have previously started new businesses. On average, the nascent entrepreneurs in the United States have between 10-15 years of work experience and just over 20 per cent have had experience starting other new businesses. There are no statistical differences between the experiences of women and men. Among the Norwegian nascent entrepreneurs, men are three times as likely as women to have tried to start a new business sometime in their life and given up (31 per cent *vs.* 13 per cent). To the extent that learning occurs through the experiences of the start-up process, Norwegian men would appear to have accumulated greater human capital useful in establishing new businesses than have Norwegian women.

Financial resources

Both systemic barriers and overt discrimination have been viewed as restricting women's access to resources for starting new businesses. Access to financial resources has been regarded as one of the most important resources denied women. At start-up the most salient issue seems to be where the money will be acquired.

In the United States men and women anticipated having different sources of financing to establish the new business. Four sources of funding were examined: *i)* loans and equity the entrepreneurs intended to contribute from personal sources; *ii)* equity and loans secured from family and friends; *iii)* equity and loans from banks and private investors; and *iv)* money contributed by other start-up team members. The data reveal that men anticipated having considerably greater access to formal sources of money (banks and private investors), and that they intended to contribute substantially more of their own money than women. Men expect that more than 47 per cent of their financing will come from outside formal sources in comparison to women who expect this type of funding to represent only 26 per cent of their total resource base. In contrast, women report that personal financial sources will make up more than half of the total financial base they will have available at start-up. Men expect only one-third of their total to come from personal equity and loans to the business. Neither women nor men anticipated relying much on family and friends or other start-up team members as sources of financing: only about 10 per cent will be contributed by each of these sources.

In addition to men intending to rely on formal sources of financing to start the new business, it also appears they have access to greater amounts of personal financial resources than do women. In the United States 45 per cent of the women report having household income of less than $30 000 per year as compared to 27 per cent of men. In Norway, 64 per cent of men starting new businesses report having household income of more than the equivalent of $50 000 as compared to only 36 per cent of the women entrepreneurs.

In general, men anticipate having substantially more money at the start-up stage than do women, they are more likely to rely on formal sources of financing and they have more personal resources in the form of household income. These findings are consistent with previous studies of established businesses. However, the finding that women do not see family and friends as viable sources of financial support differs from previous studies. This anomaly may be related to differences in the life stage being studied. Previous research has focused on established businesses. It may be that women who succeed in getting their businesses up and running can do so only with the assistance of family and friends. At the nascent stage of development, however, it may be that women do not yet view these as desirable sources. It may be that only when all other financing options are exhausted do they turn to family and friends.

Assistance programmes

The extent to which entrepreneurs successfully take their businesses through the gestation process may relate to the types of assistance they seek and receive along the way. The WI-Nascent survey asked the respondents about 28 different categories of services provided by government agencies or private sources. In addition to questions about general awareness of the programmes, the entrepreneurs were asked whether they had actually sought and received help. Table 2 displays a comparison of nascent entrepreneurs and those who, at some time, tried to get a business started, but gave up.

Discouraged women were almost three times less likely to have sought and received help than the other groups. The areas in which this difference was most striking were in their use of continuing education courses and making use of programmes designed to provide a network of contacts. Only 2 per cent of discouraged women obtained help from continuing education. In contrast, 20 per cent of discouraged men sought and received help from these courses, and over 35 per cent of nascent entrepreneurs (both women and men) took advantage of this resource. Only 18 per cent of discouraged women received assistance from networking programmes in comparison to 39 per cent of nascent women.

Table 2. Use of assistance programmes

Type of assistance	Discouraged		Nascents	
	Men %	Women %	Men %	Women %
Accounting assistance	26	25	25	31
Business advocacy groups	2	0	15	13
Business incubator	9	0	4	0
Business feasibility review	1	0	5	0
Business plan development	17	2	15	15
Business start-up or expansions	16	1	23	14
Continuing education courses	20	2	38	35
Financial counselling	23	2	23	13
Information on financial resources	23	10	18	8
General business management	9	10	14	31
Government procurement assistance	1	0	0	0
Information on laws and regulations	42	33	37	50
Information systems assistance	8	1	10	16
International trade assistance	7	1	5	0
Manufacturing systems – technical	15	0	18	1
Marketing – demographic profiles	19	16	4	13
Marketing – programme assistance	22	2	4	8
Marketing – research assistance	11	1	5	8
Minority business assistance	9	1	0	2
Networking/contacts	35	18	28	39
Patent and trademark assistance	7	0	8	12
Personnel management assistance	15	2	15	13
Retail/service business location	24	1	19	15
Site location assistance	14	1	10	0
Technology development assistance	16	0	19	0
Tourism business assistance	15	1	4	0
Venture capital	0	0	0	0
Women's business assistance	0	0	0	7
Average	**15**	**4**	**13**	**12**

Nascent women entrepreneurs were substantially more likely than the other three groups to get general business management assistance (31 per cent *vs.* approximately 10 per cent for the other groups), accounting assistance (31 per cent *vs.* approximately 25 per cent), information systems assistance (16 per cent *vs.* approximately 9 per cent), and to seek information on laws and regulations (50 per cent *vs.* approximately 35 per cent). Among the types of assistance sought by all the entrepreneurs (both women and men), the two most frequently taken advantage of were accounting assistance and information sources on laws and regulations.

One of the most curious findings was women's low reliance on programmes specially designed for them. None of the discouraged women entrepreneurs reported getting help from women's business assistance programmes, and only 7 per cent of the nascent women used such assistance. Further analysis of the data suggests this was not due to a lack of awareness. For each of the types of assistance programme, respondents were asked to indicate whether they "couldn't find help" or "didn't know about help". Both groups of women seem to be informed about the availability of women's assistance programmes. Only 7 per cent of each group indicated that they didn't know about such programmes, and none indicated that they couldn't find the help when they sought it. Thus, to increase the utilisation rate of these assistance programmes, those responsible for the programmes may need to illustrate and communicate more effectively how and why the services will benefit women.

In general, there appear to be differences in the pattern of programme utilisation by women and men, and by whether entrepreneurs became discouraged and gave up. Nascent women entrepreneurs were most successful in getting the help they sought. Discouraged women entrepreneurs sought and received the least help.

The preceding sections illustrate how the factors stimulating the propensity to select self-employment differ by gender of the founding entrepreneur. The findings reveal that women are less likely than men to engage in entrepreneurial initiatives. But for those who do, women appear to have more positive judgements about the entrepreneurial climate, they are more likely to have worked part-time or to have been underemployed prior to embarking on creating the new business, and they appear to be creating businesses that integrate their desire to create wealth for their families with their preference to remain in their local community.

Start-up activities in creation of the business

Our assumption in examining the influences of incubator and facilitative factors was that the founding of new firms is a process of activity rather than an identifiable event. As such, what entrepreneurs do to get the business established may be as important as what resources they bring to the creation process. In other words, the sequencing of start-up activities is thought to be critical to the future success of the business. Nascent entrepreneurs who are able to get a business up and running may undertake different activities (or sequences of behaviour) in starting their businesses than nascent entrepreneurs who fail to start the business. The question is what specific behaviour, or sequences of behaviour, result in the successful creation of a business? Table 3 displays a listing of the typical activities that entrepreneurs undertake in starting new businesses. Survey respondents were asked to indicate whether they had initiated or completed each activity and, if they had, when. Table 3 reports how often each activity is undertaken as the first activity

The findings reveal that the most common first activity undertaken varies across both gender and country. Among Norwegian entrepreneurs two-thirds of the respondents reported giving serious thought to the new venture as their first activity. Second in frequency was the formation of a start-up team. In the United States more than one-third of entrepreneurs reported investing their own savings as the first activity.

Table 3. First activities[1]

	Norway		United States	
	Men	Women	Men	Women
	%	%	%	%
Give serious thought	66	63	42	28
Form start-up team	40	34	22	8
Save to invest	36	37	29	16
Prepare business plan	26	8	22	12
Look for equipment/facilities	15	18	24	28
Search for customers	15	8		
Invest own savings	12	16	33	36
Market research	11	10		
Marketing programme			2	0
Product development	8	13	2	12
Government registration	7	3		
Devote full time	6	3	9	4
First sales	4	5	16	12
Buy equipment/facilities	3	8	29	12
Positive cash flow	3	3	9	4
Ask for funding			24	12
Government funding	1	5		
Private funding	1	8		
Obtain funding			9	16
Private funding	1	8		
Government funding	0	0		
Hire employees	0	0	0	0
Apply for licence or patent			9	12
Adopt legal form			4	0

1. The columns do not total to 100 per cent since activities may be initiated simultaneously.

To gain additional insight into the sequencing of the activities, we developed a time scale based on when the respondent reported that each activity had been initiated or completed. Values ranged from 1 (corresponding to activities initiated within a month of the first behaviour) to 18 (corresponding to the fifth year of start-up activity). For example, a value of 3 signified activities that were initiated during the third month after the first activity was performed. Table 4 displays the sequencing of activities among Norwegian entrepreneurs.

The findings reveal that Norwegian men and women vary somewhat in the way they sequence start-up activities. More than 50 per cent of the men in Norway begin the process of building their businesses with a written plan. Norwegian women typically don't have a business plan until eight months after they've started other activities to establish the business. Women tend to ask for and receive government funding as much as a year earlier in the start-up process than do men (four months for women *vs.* 18 months for men). Men, on the other hand, are likely to hire employees more than six months earlier than do women (nine months for men *vs.* 16 months for women). Finally, men are more apt to register their business with the government within the first six months of their start-up activities, whereas women tend not to register until they've been in the process for over 16 months.

Table 4. Norway – Start-up activities

Median months

	Men	Women
At start	Give serious thought Form start-up team Save to invest Prepare business plan	Give serious thought Form start-up team Save to invest
1-3 months	Product development Market research	Product development Market research Invest own savings
4-6 months	Look for equipment/facilities Invest own savings First sales Commit to full-time work Search for customers Government registration	Look for equipment/facilities Buy equipment/facilities Ask for government funding Receive government funding
7-9 months	Positive cash flow Buy equipment/facilities Ask for government funding Obtain private funding Hire employees	Positive cash flow First sales Commit to full-time work Search for customers Prepare business plan
10-12 months	Ask for private funding	Ask for private funding Obtain private funding
13-18 months	Obtain government funding Apply for licence, patents	Government registration Hire employees
19-24 months		Apply for licence, patents

Table 5 reports the sequencing of activities among US nascent entrepreneurs. The findings reveal that in the United States more than 50 per cent of men starting new business begin by asking for funding. Both men and women are apt to get funding within the first three months of the start-up process. Men tend to devote full-time to the start-up within the first six months of the process. Women in the United States tend not to work full-time in the business until almost a year-and-a-half after the first start-up activity. Similarly, men are likely to hire employees before women do. More than 50 per cent of the men reported hiring employees within nine months of the first start-up activity, whereas women tended not to hire employees until they'd been in the process for almost two years.

Table 5. United States – Start-up activities

Median months

	Men	Women
At start	Ask for funding	
1-3 months	Invest own savings Prepare business plan Obtain funding Save to invest Look for equipment/facilities Buy equipment/facilities	Invest own savings Prepare business plan Obtain funding Product development
4-6 months	Form start-up team Positive cash flow Devote full-time work	Look for equipment/facilities Buy equipment/facilities Ask for funding Adopt legal form Save to invest
7-9 months	First sales Product development Adopt legal form Hire employees	First sales Form start-up team Positive cash flow
10-12 months	Marketing programme Apply for licence, patents	Marketing programme
13-18 months		Devote full-time work Apply for licence, patents
19-24 months		Hire employees

A comparison of the ways in which women begin their businesses in the United States and Norway reveals similarities and differences. Women in Norway and the United States evidence similar early commitment to the new venture. More than 50 per cent of women in both countries who invest their own savings in the new venture do so within the first three months of the start-up process. Women in both countries tend to have their first sales and positive cash flow within nine months of the first start-up activity. Similarly, for those hiring employees, they tend to delay that additional overhead until a year-and-a-half to two years into the process. They also tend to delay getting into the business registries.

Cross cultural differences in the start-up process can be seen in preparation of a business plan, obtaining funding and in forming a start-up team. Women in the United States are likely to develop a business plan much earlier than do women in Norway. More than 50 per cent of the women in the United States who wrote a business plan did so within the first two months. Women in Norway tend not to prepare the plan until the eighth month of the start-up process. In the United States more than 50 per cent of the women who obtain funding do so within the first month. In Norway, 50 per cent of the women who receive government funding do so within four months of the first start-up activity, but they don't obtain private funding until after nine months. Women in Norway tend to adopt a start-up team much earlier than do women in the United States. In Norway, more than 50 per cent of women forming a start-up team do so as one of the very first start-up activities. Women in the United States tend not to take on partners until eight months after the first start-up activity.

From gestation to new business

The time period of activity from when entrepreneurs first take action to start the new venture to when the firm becomes a viable economic entity has been termed the "gestation period". The findings discussed above indicate that women are less likely than men to engage in entrepreneurial initiatives and that they bring different resources than men and adopt differing sequencing of start-up activities. But to what extent does their success in making the transition from the nascent stage to fledgling new firm status depend on those resource and sequencing differences? Data collected during follow-up interviews with the respondents from the WI-Nascent and US-Nascent surveys provide insight on this question. Erosion in the sample between the two data collection periods requires caution in interpreting the findings. However, although tentative, the findings are instructive.

Of the nascent entrepreneurs interviewed during a second data collection, only 28 per cent of women reported that their business was up and running. Of the men, 63 per cent reported the status of their firm as a fledgling new business. Just over 30 per cent of the entrepreneurs indicated that they were still continuing efforts to establish the business. Women were much more likely than men to have put their efforts on hold for the time being (a hiatus) or to have become discouraged and have given up the process. Thus, it would appear that not only do fewer women select self-employment as a career objective, but those who do have more difficulty than their male counterparts in getting the firm through the gestation process. Further research will be required to verify the extent to which these findings are a reflection of the regional context. Data from these samples will need to be compared with those of other countries and geographic regions.

Table 6 displays the sequencing of activities reported by Carter, Gartner and Reynolds (1996) for three groups of nascent entrepreneurs: those who successfully made the transition from the gestation stage of the organisation's life cycle to that of fledgling new firm; those who became discouraged between the first and second data collection periods and gave up; and those who at the time of the second interview reported that they were still attempting to get the business established.

The findings indicate that those nascent entrepreneurs who were able to start a business were more aggressive in making their businesses real; that is, they undertook activities that made their businesses tangible to others; they looked for facilities and equipment; sought and obtained financial support; formed a legal entity; organised a team; bought facilities and equipment; and devoted full-time to the business. Individuals who started businesses seemed to act with a greater level of intensity. Those who had become discouraged and given up may have discovered that their initial idea for the business would not lead to success. These entrepreneurs started by developing a model or prototype to "test" out their ideas. They may have found that the product/service would not work according to their expectations. These entrepreneurs were similar in their activity patterns to those who got their businesses started. They appear to be just as aggressive at the start but, as the business unfolded over time, they decreased their activities and then ceased start-up. Women predominate in this category.

Entrepreneurs who reported that they were still trying to put the business in place appeared to be adopting a slow deliberate pace. Their focus was internal to the start-up process, saving money and preparing a plan. Less effort was being made to make the business real to others. Women were more likely than men to be still trying to get their business established at the time of the second data collection.

Table 6. Sequencing of start up activities

Year	Started	Gave up	Still trying
1st month	Looked for facilities/equipment	Asked for funding Developed product Saved money to invest Organised start-up team	
1st quarter	Invested own money Asked for funding Obtained funding Prepared business plan	Invested own money Obtained funding	Saved money to invest
2nd quarter	Formed legal entity Bought facilities/equipment Formed start-up team Organised full-time	Prepared business plan Looked for facilities/equipment	Prepared business plan Organised start-up team
3rd quarter	Hired employees	Bought facilities/equipment	Invested own money Looked for facilities/equipment Applied for licence/patent
4th quarter	Saved money Rented facilities/equipment Applied for licence/patent	Devoted full-time Applied for licence/patent	Bought facilities/equipment
Distribution of women	**24 per cent**	**40 per cent**	**36 per cent**

Note: Categorised by median value for those who had initiated activity. Median values reported only if more than five entrepreneurs in the group had initiated activity.

Summary

The US and Norwegian studies provide data for contrasting the experiences of women and men as they attempt to establish and develop new ventures. Two sets of factors, thought to explain why the experiences of women starting new businesses may differ from those of men, were explored. *Incubator influencers* were examined to see if they differentially stimulate the "triggering" event for self-employment. *Facilitative factors* thought to differentially affect the start-up process once the initial entrepreneurial decision was made were also studied. Additionally, the activities undertaken by entrepreneurs in first establishing the business were examined to determine whether certain patterns predict greater success in putting the fledgling business in place. Among the differences isolated in the analyses are the following:

◊ The ratio of participation in the entrepreneurial process varies by gender. Norwegian men are almost three times as likely as Norwegian women to be working to establish new businesses. In the United States men starting businesses outnumber women by 2 to 1.

◊ US women entrepreneurs tended to have more positive judgements about the entrepreneurial climate than men, but thought government programmes were more supportive of men. US women also believed that more people would start new businesses if it were not so complicated to get help from the government.

◊ US women who had tried to get a business established but who had become discouraged and given up associated the greatest social stigma to failure, indicated they were least likely to try another entrepreneurial initiative, and were most dissatisfied with access to financial resources to help start their businesses.

◊ Few life-stage differences exist between men and women entrepreneurs in the early stages of creating their new business.

◊ Both Norwegian and US women are more likely than men to have been working part-time just prior to initiating start-up activities. Women's work status may indicate that they are underemployed relative to men, and self-employment may represent the most attractive alternative to part-time employment. In Norway fully 71 per cent of the respondents indicated self-employment as their preference. Yet just over 20 per cent of women in both countries report being self-employed sometime in their life.

◊ Norwegian men appear to have accumulated more human capital from attempting to start previous businesses than have Norwegian women.

◊ Like men, US women seek to create businesses that provide wealth for their families. Similarly, both women and men value remaining in their local community more than having autonomy in their job or seeking a position that has high task interest.

◊ Women entrepreneurs in Norway and the United States tend to come from households where the level of income is far less than the household income of men trying to start new businesses. Not surprisingly, women expect to have considerably fewer financial resources at start-up than men, especially from outside loans.

◊ US men tend to have higher expectations for the growth of their firms than do women entrepreneurs. Men expect to have three times the level of sales expected by women entrepreneurs by the end of the tenth year of operating their businesses.

◊ Despite their availability, US women entrepreneurs do not tend to seek or receive help from women's assistance programmes.

◊ Nascent entrepreneurs in Norway and the United States demonstrated commitment to the start-up process in varying ways. In the United States the most common first step in starting a business is investing one's own savings. In Norway, a common first action is to form a start-up team.

◊ In Norway men are far more likely than women to begin the start-up process with a written plan.

◊ Norwegian women tend to ask for and receive government funding as much as a year earlier in the start-up process than do men.

◊ In the United States, attrition during the start-up process is much higher among women than among men. Almost 60 per cent of the male respondents reported successfully getting a fledging business started during the study, compared with only 24 per cent of women.

In summary, the comparisons between women and men engaging in entrepreneurial activities suggest that they share many of the problems encountered in establishing new ventures. There are, however, some areas where experiences differ. In particular, financing may be a more significant issue for women. Women appear to have fewer financial resources available at start-up. They tend to come from households with lower income levels and they register intentions to start the firm with considerably less financial support, particularly debt. These findings may suggest that women are not as aware of the magnitude of resources required to establish a business and/or that they have less access to financial resources.

Data from the follow-up interviews reveal that the chances of US women and men making the transition from nascent entrepreneur to new firm owner vary substantially. Only 24 per cent of the US women nascent entrepreneurs succeeded in getting their firms through gestation to fledgling new firm by the time of the second interview, compared with 58 per cent of men. Women entrepreneurs were more than twice as likely to have given up on starting their businesses. Thirty-six per cent of women reported still working to put the business in place. Thus, not only is women's rate of participation in the entrepreneurial process as nascent entrepreneurs less than that of men, they also appear to have a higher attrition rate during the process. However, the sample size and, particularly, the small number of women who made the transition to ownership, call for caution in generalising the findings. Further studies are needed to verify and elaborate the results. For example, barriers previously thought to inhibit women's chances of making the transition, such as more children living at home or lower levels of human capital, received marginal support here. It may be that these factors are more discriminating among established new firms than at the early stages of the gestation process.

The findings demonstrate that what entrepreneurs do in their day-to-day activities matters. The kinds of activities that nascent entrepreneurs undertake and the sequence of these activities has a significant influence on their ability to successfully create new ventures. The findings suggest that the behaviour patterns of nascent entrepreneurs who have successfully started a new venture should be identified and differentiated from the behaviour patterns of nascent entrepreneurs who failed.

Policy implications

Findings regarding the process whereby individuals elect to become involved in the gestation of a new firm in Norway and the United States suggest a number of policy issues that justify attention:

◊ role modelling opportunities;

◊ heightening of awareness of the existance of assistance programmes;

◊ efficient access to assistance programmes;

◊ acquisition of relevant "start-up" knowledge/skills;

◊ access to financial resources.

The findings among US nascent entrepreneurs suggest that the entrepreneurial climate may play an important role in triggering women's decisions to select self-employment. Women who became discouraged during the early stages of developing the business and gave up were much more likely to view the climate as negative. To the extent that perceptions of climate can "trigger" the entrepreneurial process, programmes should be developed that fully represent the opportunities available to women. Profiles of women business owners must be more readily available and widely disseminated to facilitate role modelling behaviour.

Heightening the awareness among women of the availability of assistance programmes may be among the most important and difficult areas to pursue. Despite the myriad of programmes sponsored by government agencies, economic development entities, financial institutions, associations, etc., US women make little use of these schemes. Some of the ways to improve programme delivery might include mass-media campaigns, utility bill inserts, information contained in tax statements or other government household mailings. Perhaps entrepreneur/small firm ombudspersons could be designated to help novice entrepreneurs learn about assistance programmes. Such awareness is especially critical regarding sources of financing. Few women are aware of the financial resources provided to new firms or how they can tap into these funds.

In addition to heightened awareness, the programmes offered are often poorly organised or fragmented. Efforts should be undertaken to develop effective mechanisms for co-ordinating the programmes, perhaps by creating an ongoing inventory. This would provide guidance to those seeking help as well as to those attempting to develop new assistance programmes. Similarly, the creation of a single, centralised telephone number to assist women in locating the types of services most appropriate to their situation should be considered.

The findings in this study suggest that successfully moving through the gestation stage to creation of a fledgling business may depend to a great extent on having a start-up team. Women are much more likely to delay forming a team. This may be because they do not have access to individuals who will commit themselves to and support the venture. Policy initiatives that would influence the founding of new firms would be useful. Programmes to provide counselling and advice on how to find potential partners, assist in planning or point to other ways to acquire education useful for the start-up process should be developed. Discouraged women entrepreneurs in the United States were least likely to obtain assistance in networking or help in developing managerial/operational skills for running their businesses.

Finally, it is clear that women are disadvantaged by their lack of financial resources. Both in Norway and the United States, women nascent entrepreneurs come from lower-income households than do men. In the United States women were less likely to know about the financial resources that were available, they were pessimistic about having access to financial resources, and few had sought financial counselling. Awareness programmes, as well as financing programmes, should be developed in order to minimise the possibility of women dropping out of the start-up process during the gestation stage of the business' development.

NOTE

1. Combined data from the Wisconsin and US samples are referred to in the paper as US-Nascents.

BIBLIOGRAPHY

BATES, T. (1985), "Entrepreneur Human Capital Endowments and Minority Business Viability", *The Journal of Human Resources*, 20, pp. 540-554.

BIRLEY, S. (1986), "The Small Firm – Set at the Start", paper presented at the Babson Conference on Entrepreneurship Research.

BRUSH, C. (1992), "Research on Women Business Owners: Past Trends, a New Perspective and Future Directions", *Entrepreneurship Theory and Practice*, 16(2), pp. 5-30.

CARTER, N.M., W.B. GARTNER and P.D. REYNOLDS (1996), "Exploring Start-up Sequences", *Journal of Business Venturing*, 11(3), pp. 151-166.

CROMIE, S. and S. BIRLEY (1991), "Networking by Female Business Owners in Northern Ireland", *Journal of Business Venturing*, 7, pp. 237-251.

US HOUSE OF REPRESENTATIVES (1988), New Economic Realities: The Rise of Women Entrepreneurs: A Report of the Committee on Small Business, US Government Printing Office, Washington, DC.

Chapter 10

STRENGTHS OF WOMEN AND OPPORTUNITIES FOR SMEs IN INTERNATIONAL TRADE

by
Ingrid Andersson
President, PREFO AB, Stockholm, Sweden

The performance of women in entrepreneurial activity is influenced by their ability to turn their special features into strengths instead of weaknesses. Today there are better options for this than used to be the case, both because of new openings following from extensive regulatory reform and liberalisation of services as well as other sectors, and also because of the advancement of information and communication technology. The exploitation of this technology provides new opportunities for both men and women, but it particularly allows women to compensate for what has traditionally been their greatest weakness, and to reach out with what will remain their absolute and relative strengths in business.

Today most business women tend to focus – more so than men – on the domestic market. The reasons for this have to do with the kinds of activities in which women tend to engage as well as the characteristics that are typical of women's approach to business. In this paper, however, I argue that there currently exists great potential for women to become more successful internationally. Indeed, in today's globalising economy, success in this respect is a prerequisite if women are to catch up with men, rather than falling behind, in terms of business performance. This applies even to those economies and sectors which have so far had a predominantly domestic orientation.

The performance of women in entrepreneurial activity is influenced by their ability to turn their special features into strengths instead of weaknesses. Today there are better options for this than used to be the case, both because of new openings following from extensive regulatory reform and liberalisation of services as well as other sectors, and also because of the advancement of information and communication technology. The exploitation of this technology provides new opportunities for both men and women, but it particularly allows women to compensate for what has traditionally been their greatest weakness, and to reach out with what will remain their absolute and relative strengths in business.

In this context, special attention should be paid to the concept of *trust*. Although this concept has long been acknowledged as capturing key features of successful business and well-functioning social life in general (Arrow, 1974), its economic implications, and how its virtues can best be exploited, are too seldom addressed. I will approach the significance of this matter for women entrepreneurs in international trade by building on my personal experience in this field. Some of the arguments and conclusions will be specific to the services sector. Certain aspects will also be most relevant to the challenge of entering the Japanese market, which is characterised by some unique features. This market is of considerable general interest since it is arguably one of the most difficult for newcomers to enter. Beyond this, however, I will identify principles that can be more or less generalised. These relate to certain fundamental issues which confront women entrepreneurs in international business operations and access to foreign markets, although they take different shapes in different country settings.

The study is organised as follows. In the following section I present the specifics of my business idea and the basic features of my corporation. Potential special strengths in women's approach to international business are discussed in the third section, while the final section addresses implications for policy.

Business concept

Two fundamental concepts lay the basis for my business corporation. The first is associated with the fact that SMEs generally have considerable difficulties in reaching distant markets. For example, some 40 per cent of SMEs in the OECD area are currently sheltered from global competition (OECD, 1997). Furthermore, research indicates that the level of exports can be broadly related to three factors: sector, size and managerial attitude (Duchenaut, 1997). Firms usually start out domestically, and then approach foreign markets in a sequential fashion, beginning with the most neighbouring ones. Entering markets which are more remote, in terms of physical distance as well as institutional or cultural conditions, tends to require particularly extensive research and preparatory efforts. Many entrepreneurs feel uncomfortable, or do not have the time or resources, to even begin to try out business opportunities in markets where the culture, language, rules, regulations and practices of doing business differ markedly from home.

Many SMEs are thus unaware of possible business opportunities in distant markets. Confronted with various offers from middlemen and agents, these firms tend to become stuck in more or less serious *ad hoc* engagements. The chances of success in such arrangements are modest. If trade does take place, the traders tend to capture high margins while the payoff is limited for the SMEs in question.

My second point is that most business does not consist in one-shot transactions. As suggested by Commons (1934), trade to a large extent hinges on harmonizing relations between different counterparts, and creating a situation which allows for repeated bundles of transactions. Arrow, Williamson and others have clarified the importance of mutual "trust" as a foundation for business relationships. Noticing a special strength that women have in establishing trust, the basis of which will be further discussed in the following section, and linking it to the weakness of SMEs in distant markets, I gradually identified the opportunity for a service which could create economies of scope through a credible improvement and extension of exports by SMEs to markets which they themselves would not be in the position to address effectively.

In the early 1990s, I started out as a trading agency with special emphasis on the Japanese market. Japan was chosen due to the combination of huge market potential, presence of barriers to entry and my personal interests and experience. The core challenge was to develop a shortcut through the Japanese distribution system, which is characterised by highly special, customised relations. Two Swedish industries appeared particularly good candidates because of high product quality combined with a predominance of small firms focusing on the domestic and neighbouring European markets (except for *ad hoc* consultancy arrangements) – furniture and textiles. The former gradually proved to represent the most tenable avenue.

Regarding market segments, it was clear from the outset that commercial customers, *not* the home environment, represented the appropriate and most promising area. Office space, hotels and restaurants were the original targets but, after the bubble economy burst, these areas became less interesting because of tightened budgets and less construction. For the last few years we have concentrated our efforts on the welfare sector; focusing mainly on the rapidly expanding care systems for the elderly but also to a certain extent on institutions for education. This focus has proven quite successful.

In addition to simply undertaking trade, it has been crucial *vis-à-vis* business partners on both sides to carefully communicate the vision and strategy underlying my business, not least because of the time and patience which is required in order to succeed in Japan. In the process, the operation naturally evolved from a simple trading company into an organisation providing a wider set of services. On the domestic supply side, the task from very early on incorporated the provision of frank advice to the suppliers, medium-sized furniture manufacturers in Sweden, on how to improve and change their products in order to meet the requirements of their Japanese customers. In some cases only minor alterations had to be made, while in others it was necessary to create a completely new "Japan-ised line". This adaptation to customer needs is not a one-time exercise, but rather a continuous process which never stops. Few of the suppliers which were involved at some point turned out to be capable, or willing, to make the repeated efforts that were needed and to accept the associated risks. Confrontation with the demanding Japanese customers has, however, proved to be a highly fruitful experience for the few furniture manufacturers that have stood the test. It has led them onto a route of generally increased awareness of customer satisfaction and strengthened capacity for product improvement and quality control. This has helped the few responsive companies to boost their exports to other markets as well.

At the same time, we have developed the capacity to offer a full package of services *vis-à-vis* our Japanese clients. In addition to the products themselves, this includes counselling on interior design, drawings, colour schemes, after-sales service, consultancy service and the arrangement of study visits to Scandinavia.

Today, the key characteristics of the operation can be summed up as high-level service and flexibility. In addition, given the time and uncertainty that characterise the behaviour of most Japanese purchasers, limited risk-taking is a must. I have maintained limited scale, avoided dependency on credit financing and relied on a small core of key staff, while focusing the business operation on a network of the most trustworthy external suppliers, customers and partners. The company has also established a flexible and well-adapted network of architects, staff of welfare institutions (doctors, nurses and care workers), designers, distributors and construction companies both in Sweden and in Japan. Several members of our group are self-employed female entrepreneurs, who are contracted as consultants for various projects. While the relationships are of a long-term nature, some of these partnerships are activated on a temporary basis in connection with specific projects. The flexibility of the organisation makes it possible, for example, to quickly switch between market segments as well as to adapt to new circumstances in other ways.

The internationally oriented woman

The above section indicates that there may be openings for turning the gender dimension into an advantage, while looking at this in connection with other determinants of success in international business. To address this aspect further, we need to consider not only whether and how women differ from men in their approach to international business but, equally important, whether they are perceived and treated differently. Let us first consider some female strengths. Educated women, at least in Europe, generally have greater knowledge of foreign languages than their male counterparts, reflecting the relatively greater emphasis by women on human sciences in university studies. As can be seen from Table 1, on average about half of female university graduates in OECD countries were classified in human sciences as of 1994, while only 5 per cent belonged in engineering and architecture science and 3 per cent in mathematics and computer sciences. The corresponding figures for male graduates were 27 per cent in human sciences, 23 per cent in engineering and architecture sciences and 6 per cent in mathematics and computer sciences. These patterns originate from fundamental differences in expectations and choices made early in life, although research also indicates that the two sexes may display a different genetic disposition for various capabilities.

Furthermore, the typical profile of female entrepreneurs involves characteristics such as modesty and a less pronounced profit orientation; traits which actually put up barriers for obtaining finance. In fact, it has been observed in many studies that women tend to have relatively important objectives for their operation other than generating commercial profits (NUTEK, 1997). In international business, this kind of profile, or the anticipation of it, can be useful, if cherished and exploited correctly. In my own case, for instance, my Japanese counterparts appreciate my patience and ability to listen. A "male-oriented" hunger for quick profits, no doubt especially prevalent in some cultures, is regarded as short-sighted and untrustworthy in Japan – and probably in many other societies as well.

Table 1. University graduates by subject, OECD countries, 1994, country average

Subject	Medical sciences	Natural sciences	Mathematics and computer sciences	Human sciences	Law and business	Engineering and architecture
Men	8.4	11.1	6.2	26.7	25.2	23.0
Women	13.6	8.1	2.7	49.4	21.1	5.4

Source: OECD Indicators, 1996, Table 14.3.

Whether such perceptions do or do not emanate from the behaviour or the individual, I argue that female entrepreneurs *can* enjoy an advantage in building trust in business relations. Openness in this respect can, of course, also turn into a disadvantage if it is applied in the wrong context. Many entrepreneurs are too trusting at an early stage and get bogged down in unproductive partnerships. It is probably correct to say that women should be particularly careful not to trust their counterparts on insufficient grounds, so as not to find themselves stuck with the wrong partner. If they can master this initial difficulty, however, women entrepreneurs can achieve special gains, at least under certain circumstances that I will return to below.

These traits help explaining why women-owned SMEs, as shown in Professor Ducheneaut's report, engage in services to a much greater extent than manufacturing. Of course, it should be kept in mind that today the boundaries between manufacturing and services are becoming increasingly blurred, and that services are often becoming crucial for success in manufacturing as well. Broadly speaking, however, international trade has so far remained far less developed in services than in manufacturing, *e.g.* because of the greater difficulties involved in storing and transfering services compared to goods, and also because of greater policy-induced barriers to trade in services. With the advancement of information technology, however, there is a revolution in the preconditions for trade in services, and national as well as international regulatory frameworks are in the process of opening up in order to capture the new opportunities. The position of women in this area, the still untapped potential for services combined with the rapid expansion of the services sector, together put in place a crucial challenge for female entrepreneurs to go international.

Recently there have been many discussions about how to involve women in traditionally male-dominated industries. For my own organisation it can be said that I do business in a fairly female-oriented range of sectors, which is how many people would characterise interior design, furniture and not least our specialisation on the welfare system. What is then meant by a "male" or a "female" sector? A difference has already been noted between the industries in which male or female entrepreneurs choose to do business. Of all self-employed women, some three-quarters engage in sales and services throughout the OECD countries, compared to just a little over half in the case of their male counterparts (ILO, 1995). Again, this reflects the basic differences between the sexes, their relative skills and approaches in human relations, and how these relate to the nature of different activities. Business in so-called male-dominated sectors, basic industries – steel, oil, timber – or engineering, is rooted in a traditional society which was dominated by men, and is characterised by technologies that women generally have had limited exposure to (see Table 1). On the other hand, there are also female-dominated sectors, such as care, cleaning, cosmetics, fashion and textiles. These are closely related to female traditions, and also strengths in communication and language.

Although there is a difference in the tendency of male and female entrepreneurs to enter such different industries, there will generally be a presence of representatives for both sexes in both kinds of activities. There is a notable difference across markets (societies), however. For instance, Hofstede (1991) views "emphasis on masculine *vis-à-vis* feminine values" as one of the key dimensions in accordance with which countries can be classified. He argues that the masculine society, among other things, favours competition, individualism and success, while the feminine one cherishes equality, caring for others and consensus. According to this classification, Japan is ranked as the most masculine country, while Austria, Germany and the Anglo-Saxon countries also score rather high in this respect. On the feminine side, Sweden is listed number one, followed by the other Nordic countries, the Netherlands and some Latin American countries. In any rate, certain markets are *de facto* characterised by a relatively strong dominance by men in business, irrespective of sectors. In comparison, others are relatively more dominated by women, in practice meaning more of an equal presence, or influence, of the two sexes. A male-dominated market thus tends to have a dominance of men even in a "female" sector. A mixed market is likely to have at least a mixed influence in female sectors, but male sectors tend to remain very much dominated by men.

Given these broad differences, let us consider the opportunities for a "typical" woman entrepreneur to establish successful business relations in a certain sector/market, which is illustrated in Figure 1. Consider first the case of a traditionally "female" industry. In the mixed market the woman entrepreneur will be perceived in a neutral way, without any pros or cons (based on the fact that she is a "typical" woman). This is represented by (0) in the upper left-hand box of Figure 1. In a market dominated by males, on the other hand, she *can* actually benefit from being a woman. Why is this the case? First, women are rare in the business field of this market but, at the same time, the female entrepreneur need not be viewed as out of context, or as lacking knowledge, because of her association with the female sector. While this makes her unusual she is also potentially an asset – resulting in an openness to place confidence in her. In addition, because women are generally not competing with men on the same terms, the woman entrepreneur may be viewed as less of a threat in this case, which improves the chances for the establishment of mutual trust. Taken together, this implies that women entrepreneurs who specialise in a female-dominated sector may have a real advantage in penetrating a male-oriented market, which accounts for (+) in the lower left-hand box in Figure 1.

Figure 1. Returns to women characteristics, different sectors and markets

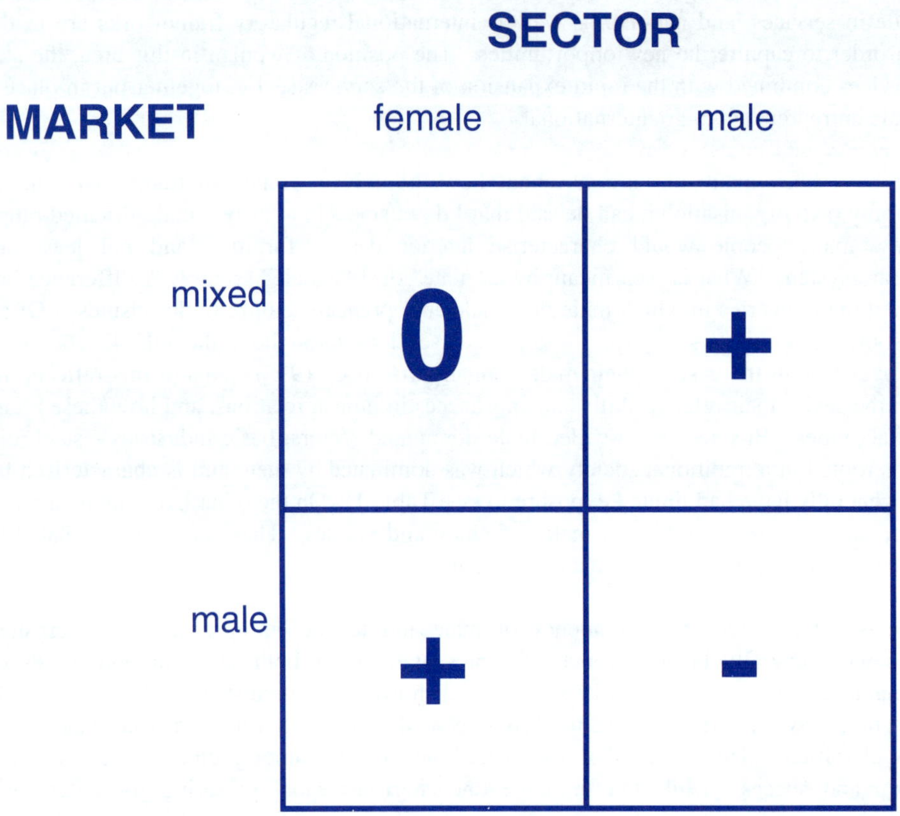

A woman entrepreneur who engages in a traditionally male industry, on the other hand, is likely to enjoy an advantage in a mixed market (+), because here again there is an openness towards women entrepreneurs, and also because she is unusual in this kind of setting. In this kind of sector, she may experience a disadvantage in a male-oriented market (-), because as a woman she runs the risk of being viewed as out of context without any natural foundation for radiating competence, resulting in weak prospects for her to communicate and establish trust with the male incumbents.

To conclude, in order to be internationally successful, there are two distinct avenues of approach for the women entrepreneur, the applicability of which is influenced by sectoral characteristics. Taking everything else as given, in a male-dominated sector, a mixed market offers the greatest advantages for the women entrepreneur, whereas in a female sector she is confronted with the relatively best conditions in the male market. Viewed from the other angle, the mixed market is the most penetrable for the female entrepreneur in a male sector, while in the male market she will be in the strongest position in the female sector. Thus, a male-dominated market may in fact not pose the greatest difficulties for a female entrepreneur in international business, although the male niches in such markets are likely to be excessively difficult.

Obviously, the above is a highly condensed illustration which should not be taken too literally. In practice, there are no such sharp dividing lines neither between sectors nor between markets. Values and attitudes also change over time, although this by no means happens overnight. Broadly speaking, however, this simplified set-up explains some stylised facts which are of considerable importance in the real world. The point is not to encourage women to do business in certain activities or discourage them from others, but rather to demonstrate the importance for women of understanding when and how to use whatever special strengths they have. In going international, women entrepreneurs need to adapt in order to exploit the advantages or deal with the difficulties which they encounter. Which approaches are the most suitable and useful, and which markets are the most accessible, vary because female features play a different role under different circumstances.

Meanwhile, it has been noted that services, in which most women specialise, have been marked by a rapid expansion of new opportunities and working methods. Companies around the world need to be aware of the new possibilities, upgrade their ability to reach out to new customers and find innovative ways of exploiting new technologies. However, many female-owned SMEs have still not grasped that they operate in the fastest growing industry and that new opportunities have arrived. For example, the emerging means for communication make it possible to compensate for factors such as limited size and remote location. With technologies becoming much more user-friendly because of competition and the importance of offering good use of customers' time, women are attaining an opportunity to overcome their traditional weakness in technical matters.

Policy implications

The above has implications for policy makers concerned with the opportunities for women entrepreneurs in international trade. Internationalisation is impossible without knowledge and information about possibilities in foreign markets. The OECD Conference on "Women Entrepreneurs in Small and Medium Enterprises" is a good example of what is needed. It is important to develop and diffuse knowledge about the potential strengths of women entrepreneurs, and above all for women and men around the world to grasp that new opportunities for mutually rewarding trade are arising.

There are already existing fora for business women: several countries have organisations where female entrepreneurs and executives meet at the local or national level. Such arrangements are indeed helpful for exchanging information and experience. However, women's participation in networks connected to international trade, whether among each other or with men, is less common. It is clear that much remains to be done at this level. Upgrading of skills for optimal use of new technologies is essential and can be combined with networking. There is one big obstacle, however: most female entrepreneurs cannot find the time required for engaging in such networking – the time they have available tends to be even more scarce than in the case of male entrepreneurs. Running a small company is *very* time-consuming, and women generally have particularly tough additional demands on their time emanating from home, family, children, and so forth. Joining meetings where the payoff is uncertain, and most likely would take a while before showing up, must often of necessity be viewed as a matter of low priority. It is also a common entrepreneurial characteristic to be somewhat of a "loner", which may be further pronounced by a fear of revealing business secrets, since many entrepreneurs also are innovators in some sense.

Against this background there is a great need for establishing more attractive fora where women entrepreneurs can gather, exchange information and help each other develop and grow. When it comes to international trade one naturally thinks of the usually public-sector export business councils. In some countries, these bodies are highly involved in assisting SMEs. It is common that efforts are made to "marry" attractive partner firms in different markets. In other cases, councils concentrate their efforts on large corporations, especially multinationals, particularly when the councils are pushed to arrange most of their financing themselves. It would be valuable to obtain a proper international evaluation and comparison of governmental export promotion programmes for SMEs, and how they are affected by the governmental preconditions for their operations.

It does appear that export councils generally spend insufficient time and effort in building up expertise which is relevant for SMEs, and this is true especially in connection with women-owned SMEs. These firms which are relatively small and operate in sectors which are growing fast and internationalising rapidly are in the greatest need of external counselling and network building. At the same time, women are known to make relatively little use of external advisory services and information, partly due to inadequate or non-existing information tailored to their specific areas and needs (NUTEK, 1997). Moreover, the establishment of business networks, the provision of information about technical and market opportunities and how to address them, do not really provide returns that can easily be captured by private consultants. Thus, governments have a role to play in this area, although women entrepreneurs should eventuallybenefit themselves by making the effort to join existing, or create their own, networks in international trade.

Finally, some governments offer "start-your-own-enterprise" courses for unemployed people. This endeavour has limited prospects, since the entrepreneurial spirit generally requires inspiration/education early in life. However, entrepreneurs who have created their own companies are often in great need of advice and training in the initial stages in order to become aware of their weak points as well as their potential strengths. Again, limited knowledge of opportunities in international trade tends to be one of their weaknesses, although men and women have different comparative advantages in this area. If more SMEs are to survive and grow in the globalising economy, companies may need to be approached on a systematic basis soon after their establishment.

BIBLIOGRAPHY

ARROW, K.J. (1974), *The Limits of Organisation*, W.W. Norton, New York.

COMMONS, J.R. (1934), *Institutional Economics*, University of Wisconsin Press, Madison.

DUCHENEAUT, B. (1997), *Women Entrepreneurs in SMEs*, Background Report to the OECD Conference.

HOFSTEDE, G. (1991), *Cultures and Organisations – Software of the Mind*, McGraw Hill.

INTERNATIONAL LABOUR ORGANISATION (1995), *Yearbook of Labour Statistics*, Geneva.

NUTEK (1997), "Women in SMEs", in *European Observatory for SMEs*, 4th Annual Report.

OECD (1997), *Globalisation and Small and Medium Enterprises (SMEs)*, Paris.

Chapter 11

WOMEN-OWNED BUSINESS AND ACCESS TO MARKETS

by
Anthony J. DeLuca
Director, USAF Office of Small and Disadvantaged Business Utilisation,
United States Air Force

The Office of Small and Disadvantaged Business Utilisation for the United States Air Force is committed to the enhanced participation of Women-owned Business (WOB) in Air Force programmes. This commitment began with the realisation that to effectively institute a viable programme to increase utilisation of WOBs in Air Force contracts, it was necessary to understand the environment in which these businesses operate and discover what type of assistance WOBs require – rather than offering what we thought was needed.

The initial improvements made to our small business programme to enhance WOB participation resulted in a 25 per cent growth over two years. However, I am of the opinion that more can be done and that additional research into current federal and industry practices could yield ideas for improvement. ... Many federal programmes exist to aid women-owned businesses. However, WOB experience with these programmes is very limited due to lack of knowledge. This indicates that any programme aimed at assisting WOBs must be advertised and marketed by the agency providing the assistance. Industry best practices underscored the need for a proactive WOB programme, visibly supported by top management, and that included discreet targets/goals for WOB awards, participation in outreach activities, membership in national/local women's business organisations, internal buyer training and merit pay tied to goal accomplishment. Our findings concluded that there were few gems or nuggets we were not already using. This leads me to believe that like quality and kaizen, improvement, to be sustained, must come from the people involved and their willingness to stay the course and recognise that the journey has only just begun.

As the Director of the Office of Small and Disadvantaged Business Utilisation for the United States Air Force, I am committed to the enhanced participation of Women-owned Business (WOB) in Air Force programmes. This commitment began with the realisation that to effectively institute a viable programme to increase utilisation of WOBs in Air Force contracts, it was necessary to understand the environment in which these businesses operate and discover what type of assistance WOBs require – rather than offering what *we* thought was needed. To accomplish this, I commissioned the Air Force WOB study on 4 November 1991. The purpose of the study was four-fold:

◊ to analyse the participation rate of WOB in USAF contracts by dollar value and type of business;

◊ to identify successes and failures with their corresponding rationale;

◊ to evaluate the services provided by WOB;

◊ to outline initiative(s) which could be implemented to enhance utilisation.

The study has proven to be an excellent springboard to advance initiatives and protocols which have demonstrated very positive results. The key, however is our ability to execute. This remains, in my view, the single most important element of any successful strategy and, often, the most demanding and difficult.

Access to capital *vs.* procurement opportunities

There is very little question that capital is required to begin and grow a business enterprise. Indeed, our WOB study, along with many others, identified access to capital as a major issue for WOB. Yet, although this access is needed, the presence of *bona fide* opportunities for WOB participation is what makes their businesses thrive and prosper. Further, as an Agency whose mission is the defence of the United States, we must continually ensure that our initiatives are relevant to this need and that our actions fit within the framework of what we are asked to do. Specifically, it is the understanding of what we are and what we can do that has made our actions to date successful. In other words, we are not a financial institution, nor should we attempt to entertain any such notion. We are, however, a consumer who has a large appetite and some unique requirements. The better our suppliers understand that need, and the more they are equipped with the understanding and awareness of our buying practices, the greater the likelihood their participation will be enhanced.

USAF WOB study leads to execution

Several recommendations emanated from our study. Each was evaluated to determine its usefulness and its practicality for implementation. Despite some who felt strongly that we should introduce a discreet set-aside for WOB, I chose not to pursue this as I did not foresee it as being practical given the political climate at the time. Indeed, we are now reviewing the viability of all set-asides. The recommendations we did adopt and their subsequent refinements can be incorporated by any buying activity that seeks to improve their processes for contracting with WOB. Further, any WOB seeking to do business with a purchasing activity could follow the guidelines we have developed. Again, the extent of success depends upon the depth and breadth of execution.

Recommendations lead to enhancement and our "Rule of one"

There were three key areas in which we chose to act: emphasizing WOB participation; standardizing procedures; and training. In order to emphasize WOB *participation* we established a separate quantitative and qualitative WOB goal in our small-business strategic plan which made increasing WOB participation and tracking an Air-Force-wide requirement. To help ensure WOBs were provided information on pending procurements, we introduced our "Rule of one" – the concept that each of our buying activities would solicit at least one WOB for every competitive solicitation above the Simplified Acquisition Threshold (SAT). Further, we began providing large businesses with WOB listings for subcontracting opportunities.

To aid in the *standardization* process, we provided WOBs with access to our contracting activities through our electronic bulletin board, which is now available on the Internet along with a host of other topics of interest. Our homepage address is http://www.safsb.hq.af.mil.

Once in our homepage WOBs are able to access our Long Range Acquisition Estimate (LRAE) which is aimed at helping businesses focus on potential new business and prospective niche opportunities. The more information we can provide in a timely fashion, the greater the opportunities for WOB participation. Further, availing ourselves of current electronic technology addresses a very real concern of time needed to understand requirements and their potential for future business.

An area which has repeatedly surfaced as one which needs continual attention is *training*. This is for both our buyers and the WOBs. For our buyers this translates into increased sensitivity towards WOBs. For the businesses themselves, this comes in the form of "how to do business". To achieve the latter, the Air Force has formed an Air Force Business Education Team (AFBET), comprised of expert small-business specialists who conduct real life "how-to" seminars throughout the United States. The response to date has been phenomenal – proving that education, training, and awareness are pivotal to WOB success.

To guarantee that WOB issues remained visible and focused, I established the position of Air Force Women-owned Business Advocate. This position reports to me and is empowered to oversee all WOB activities for the Air Force. I have also committed funds to support WOB technical assistance.

Finally, to ensure that we would not lose sight of our customers' needs and that we would remain sensitive to their situation, I formed the Air Force Women-owned Business Council. This group is comprised of WOB owners as well as representatives of large prime contractors. The purpose of the Council is to advise the Air Force on WOB strategy and be a source of innovation and policy assessment.

As evidenced by the above, there are three core ingredients necessary for successful implementation of any process improvements – planning, resources and execution.

Benchmarking for continuous improvement

The initial improvements made to our small business programme to enhance WOB participation resulted in a 25 per cent growth over two years. However, I am of the opinion that more can be done and that additional research into current federal and industry practices could yield ideas for improvement. Consequently, I launched a study aimed at identifying and assessing the effectiveness of current federal programmes and a separate study to benchmark industry practices. The results of these efforts revealed that many federal programmes exist to aid WOBs. However, WOB experience with these programmes was very limited due to lack of knowledge. This indicates that any programme aimed at assisting WOBs must be advertised and marketed by the agency providing the assistance. Industry best practices underscored the need for a proactive WOB programme, visibly supported by top management, and that included discreet targets/goals for WOB awards, participation in outreach activities, membership in national/local women's business organisations, internal buyer training and merit pay tied to goal accomplishment. Our findings concluded that there were few gems or nuggets we were not already using. This leads me to believe that like quality and kaizen, improvement, to be sustained, must come from the people involved and their willingness to stay the course and recognise that the journey has only just begun. As I have previously stated, adoption of the techniques we have developed will greatly support an activity in enhancing WOB participation. Yet, any agency or activity programme will be less than successful if the WOBs themselves fail to adopt certain key tenets.

When opportunity knocks

There are many people who observe successful entities and attribute that success to luck. Some would say that luck is the joining of preparation and opportunity. Although I believe that this is true, I am inclined to believe that success is more often the *will to be prepared* when opportunity strikes. It is often the lack of this will that causes businesses to fail.

I would now like to share with you some of the tenets I believe WOBs should pursue when attempting to capture market share:

◊ *Know your customer.* This may sound trite but it is basic. It includes having the necessary infrastructure to support their requirements.

◊ *Know your competition.* This means knowing their numbers as well as their past experience with your customer. Don't expect instant success if you are in an industry that has relatively easy access and has many like providers. To niche is always better.

◊ *Network.* Don't underestimate the power of teams.

◊ *Use the electronic medium to help focus your marketing strategy*, but remember there is nothing like personal contact.

◊ *Learn about and use the varied federal programmes available.* Surfing the Internet will assist you in this.

◊ If you want to *pursue innovation*, consider submitting a proposal under the aegis of the Small Business Innovation Research (SBIR) Program or the Small Business Technology Transfer Pilot Programs (STTR). Once again, information on these programmes is available on the Internet.

International diversity

The global opportunities that may present themselves also bring with them some unique challenges. To assist WOBs who wish to pursue international markets, I initiated a study to identify some of the requirements and challenges that would be faced by businesses entering that market. Businesses should recognise that getting started in exporting can involve significant investments in both time and money. Further, depending upon your export strategy, you may need to familiarise yourself with the types of assistance available, research and identify appropriate markets for your product/service, travel to potential markets and possibly modify your product/service for a new market. Although there are difficulties which you should be aware of, the rewards for a diversified company are attractive.

Our study identified some steps that should be undertaken when pursuing a foreign market. These include: preparing an export plan, finding market research assistance, researching and choosing a method of exporting, understanding methods of payment, understanding regulations and obtaining licences, documenting and transporting your product/service, and servicing the product/service after award.

Additional insight into our study is available on the Air Force Small Business Homepage.

Flexible methods for meeting WOB needs

Irrespective of the type of business WOBs intend to pursue, envisioning the right strategies for business assistance programmes will require more creativity and courage than ever before. To find new ways of meeting the needs of WOBs – some of which are not yet known – will require a dedication to ever more flexible methods of programme design and implementation. Some methods are offered below.

Teach electronic commerce

Ensure broad access and education for business owners and procurement personnel. As actual physical access increases and mandates for implementation are put in place, broader use of the technology should follow. A vital part of adequate access is appropriate training.

Support innovative business relationships

Support new forms of business arrangements and mentoring relationships by adopting new methods for procurement. Studies show that companies graduating from incubator arrangements have a better than 80 per cent chance of surviving and thriving.

Publish information about access to capital

Support banking industry reforms and innovative investment programmes which could enhance small-business venture capital efforts. A careful examination of common credit scoring activities is needed in order to discern the benefit or harm of this practice to small businesses.

Support procurement reform

Strengthen prompt payment regulations with heavier consequences for agencies which fail to pay small businesses on time. Investigate use of controlled second- and third-tier subcontracting programmes to provide the mentoring, quality assurance, marketing and quick-pay arrangements needed by tiny, specialised suppliers.

Demand certification and registration

The 1996 Women's Economic Summit Work Groups voiced broad requests of both government agencies and private industry: to provide and demand a national certification and one-time procurement registration for women-owned businesses.

Support data collection and research on women-owned businesses

Few procurement statistics on women-owned businesses are available. Greater access and cross-referencing of census statistics is imperative for future studies.

Co-ordination and co-operation with intermediary organisations

Provide marketing and other support for indispensable services, such as business plan design, loan prequalification and bid preparation, provided by intermediary organisations.

These recommendations are proposed as areas in which to seek more information and to continue exploring as the market-place changes. They are listed here as suggestions for more flexible concepts, which may be able to accommodate expected changes in the federal system and in small business management.

Towards a new frontier

As we look to the future, we need to recognise that Women-owned Businesses will continue to grow both in number and in type. The efforts we put forth to enhance WOB utilisation will underscore our commitment and our vision. To that end, I offer the following thoughts for consideration:

- ◊ An international consortium should be formed to serve as a "think-tank" for improving the utilisation of WOBs world-wide. Quality is about sharing to improve the whole.
- ◊ Government agencies and large corporations must commit resources to enhancing WOB participation. We have not yet achieved steady-state.
- ◊ Any off-set requirements should include opportunities for WOBs, and conversely be sensitive to potential WOB displacement.
- ◊ Adopt an international mentor-protégé programme that would facilitate global expansion of WOB opportunities.
- ◊ Embark upon a comprehensive education and outreach programme that would address internal agency/company attitudes and provide insights to WOBs on how to do business.
- ◊ Adopt the US WOB Summit '96 proposal to achieve WOB participation commensurate with their population by the year 2000.

The sharing of ideas in support of a common objective is a powerful means to achieve that end. However, the ideas that will flow from this conference will remain just that – ideas – unless we put in place the means and the wherewithal for their implementation.

Chapter 12

FUNDING FOR WOMEN ENTREPRENEURS: A REAL – THOUGH DISPUTED – PROBLEM

by
Patricia Mahot
Associate Director of Studio *Entreprendre*
Paris, France

To discuss the problem of funding for enterprises created and run by women is to implicitly recognise that there is a gender difference in ease of access to funding.

... Notwithstanding the features that women have in common with men, all the studies show that there are areas in which women are particularly hampered: difficulties in getting together start-up capital and in obtaining access to financing are the main problems cited by economic actors in contact with women wishing to set up a business.

...By exploring the access to financing of small businesses created or run by women – businesses that face special difficulties over and above the usual structural problems faced by small firms – we are helping to find solutions for all small businesses.

To discuss the problem of funding for enterprises created and run by women is to implicitly recognise that there is a gender difference in ease of access to funding. That difference is the subject of this paper. What conclusions can be drawn?

In the retail and service sectors women face the same difficulties as men

In the vast majority of cases, the businesses which women set up and run are small. For example, in France, only 7 per cent of the 30 per cent of businesses started by women have more than ten employees.[1] Women set up businesses primarily in the retail and service sectors.

Several factors are conducive to the concentration of women-owned businesses in these sectors:

◊ Women arrive later than men on the labour market and tend to be employed in low-skilled jobs. Few have any experience of management or running a business.

◊ Because of the cultural factors that shape their choice of studies and qualifications and the fact that they work in a narrower range of occupations – and which tend to be less technical than male occupations – there are whole areas of the economy in which there are very few women.

These characteristics are also found among women starting their own business; a joint study by ANCE and INSEE in France showed that the majority of women starting their own business had previously been white-collar workers in the tertiary sector, or were inactive. Furthermore, when they do have some professional experience, on average it is shorter than that of men.

Women-owned businesses encounter the same difficulties encountered by all small businesses in these sectors. Such businesses are particularly vulnerable because of:

◊ their financial situation – the lack start-up capital and their financial fragility makes it difficult for them to borrow funds;

◊ their sensitivity to market fluctuations – a downturn in the economy hits them badly;

◊ the profile of their owners – they are often alone and ill-prepared to be the head of a company.

Banks are reluctant to lend to such businesses, regardless of whether the funds are for starting a business or expanding an existing one. Often under-capitalised, the firms lack visibility, their growth prospects are low and they tend to be in sectors where the company's main asset is the entrepreneur. In such circumstances, the banker' risk evaluation is often a gamble.

A study commissioned in 1995 by the French Ministry of Industry,[2] which compared the situation in France with that of the United Kingdom, Germany, Spain and Italy, examined the following question: Are business failures directly linked to credit availability? The study reached the following conclusions:

◊ there is no scarcity of credit, rather the problem is one of distributing it to small businesses;

◊ expensive guarantees and loan administration costs push up the rates at which these businesses can borrow;

◊ the banking system does not know how to respond to the diversity of small businesses in Europe, which in terms of size and sector of activity constitute a disparate population;

◊ finally, the cost of business failures is very high for the community: on average, in Europe a new business creates 1.7 jobs, while a business that fails destroys 3.2 jobs.

Characteristics of funding for women-owned enterprises

Notwithstanding the features that women have in common with men, all the studies show that there are areas in which women are particularly hampered: difficulties in getting together start-up capital and in obtaining access to financing are the main problems cited by economic actors in contact with women wishing to set up a business. If this is the case, why is it so?

First, it should be said that the world of women entrepreneurs is not an homogeneous one. It comprises at least two main types of entrepreneurs:

◊ Those who do not have any real occupation and for whom the only solution is start their own business, and those who have an occupation and professional experience which they exploit in a firm providing services to enterprises. They tend to start businesses in the tertiary sector, a sector which is often precarious.

The size of these small businesses is a state and not a stage of development. Women are less likely than men to start a business as an incorporated undertaking; they tend rather to start it in their own name.

◊ Those who come from a background of entrepreneurs, who start a business by choice, or who take over a family business. The firms they set up tend to be in retailing, with a fairly high rate of success.

They have a strategy of winning markets. They create new value-added services or products. These small businesses have the potential to grow.

Running through all these businesses is a common thread of characteristics shared by a majority by women. Do these characteristics constitute a stereotype or a state of fact resulting from a combination of factors which have not been systematically investigated?

◊ Women tend to take less risks and to apply less for loans than men.

- Studies carried out in the Netherlands, Germany and the United Kingdom[3] have shown that women start businesses with smaller loans than men. Whether they are starting or taking over a business, the capital they put into it will be smaller than for men (ANCE/INSEE, 1996).

- They have less recourse to bank finance and prefer to rely on their own resources or to call upon their family or friends (a study carried out in the United Kingdom in 1993[4] showed that 55 per cent of men had recourse to loans compared with 33 per cent of women).

◊ They have difficulty presenting business plans that reassure the banks.

- With less experience of running a business than men, and tending to start firms in lines of business for which their qualifications are less suited, with smaller firms and in local markets, most women entrepreneurs are unable to present a business forecast.

- They may find it even harder to win over the banks in that, as the European Observatory for SMEs points out in its 1996 report, the survival rate for firms set up by women is slightly lower than that for men, partly due to the lines of business they choose.

◊ Also, in addition to running the business women often still have to run a family (lack of time, little incentive to take risks).

◊ Not to mention behavioural differences between men and women – the latter invest more in relationships, look for solutions rather than management models and run their business in a pragmatic and "reasonable" way rather than through drawing up financial projections.

These factors are amplified by a frequent motive for women starting their own business – to create their own job.

Gender differences in training, occupations and management practices are very much present in bankers' minds as well as in all strata of society, and help to fuel prejudices.

All these data tend to establish that the differences between women and men have less to do with access to financing *per se* – all SMEs whether they be run by men or women face the same difficulties – than with the fact that women entrepreneurs replicate in their business the situations they have known or experienced (type of job, lower skills and less professional and financial ambition than men, tendency to underestimate the financial resources needed to start a business, etc.).

The differences between male and female entrepreneurs are practically the same as those between men and women in the labour market. Professional identity and related cultural behaviour are factors that are crucial to an understanding of the issues. Whereas in most EU countries the level of initial training of women entrepreneurs is equivalent to or higher than that of men, they are often less competent than men in the financial and management areas.

The three main funding needs of small businesses, and the difficulties women have in meeting them

Equity

Equity is the foundation of any business. It ensures that the firm can remain financially independent of banks and enables it to borrow.

In recent years the increase in the number of new enterprises, especially those created by women, has often been prompted by the desire to create one's own job: such businesses, started as a matter of urgency in order to earn a living, very often lack capital.

Another factor which explains in part why women find it very difficult to get together the capital needed to start a business is that they were paid less than men in the jobs they held previously.

To borrow, one must have savings. That being so, how can the capital base of small enterprises be strengthened? The capital base of such businesses can be constituted in several ways, as summarised below:

a) • savings
• contributions from partners
• personal loans on honour

b) • investments by outside investors (venture capital supplied by investors from the formal or informal sector)

quasi-equity
• subsidies, personal or capital subsidies
• equity loans

Venture capital supplied by the formal sector

Venture capitalists are reluctant to invest in small businesses in the retail or service sectors unless they think they have the potential to become successful, given the difficulty involved in pulling out of such investments. Furthermore, experience shows that venture capitalists do not invest in small businesses, since the kind of money that venture capitalists want to invest is much too large for the needs of a small business.

Capital supplied by investors from the informal sector

The type of financing provided by the "business angels" of the Anglo-Saxon countries is more suited to the needs of small businesses. Investors do not focus solely on the return on investment but also take into account the extent to which it helps the business to develop and survive, as well as getting involved directly in the management. It also has to be borne in mind that many of the businesses set up by women do not really lend themselves to equity stakes by outside investors.

In addition, most entrepreneurs want to remain independent and are reluctant to let in outside investors.

For the past 15 years, various, often pioneering, initiatives have tried to meet the basic funding requirements of small businesses – for example, CLEFES in France, *Goldrausch* in Germany. CLEFES provides a private legal framework for setting up savings groups, the funds from which are lent to women wishing to start a business, selected by the group of savers. In order to expand, this scheme needs to be strengthened by partnerships with local banks. *Goldrausch* provides non-interest-bearing loans to potential women entrepreneurs whose business plans do not meet traditional banking criteria.

Some countries have put into place programmes to help women constitute start-up capital. In Europe alone, we might mention the LEO (Local Employment Initiative) programme – which has not been renewed – which from 1986-87 provided subsidies to innovative businesses to help them expand or to create jobs for women. This direct subsidy, which was provided by the European Commission, may be regarded as a form of quasi-equity.

The problem of equity is difficult to resolve. It has to be recognised that the various schemes put in place have had only a limited impact.[5]

Loan financing

Credit is still the primary source of small-business financing.

Even so, such loans, with thin spreads and high risk, are of little interest to banks – especially during the start-up phase. This constitutes a disincentive to lending. Transaction costs are not proportionate to the amounts lent, and are thus relatively greater for small sums than for large ones. As a result, small-time borrowers are charged much higher rates of interest.

However, as big corporations are now able to tap into financial markets directly, and medium-sized companies are doing little capital spending (and therefore applying for fewer loans), banks are beginning to need the business of small firms.

Admittedly, the shift is very slight and the loans are often subject to conditions (*e.g.* they must be for a local business), and in most cases the lenders tend to be large nation-wide banks rather than small institutions, which, on the contrary, are becoming increasingly selective.

In practice, this is giving rise to a segmented marketing approach, the objective of which is to tailor bank services more closely to the needs of business customers. This pattern can be observed across most European countries.

The example, in the Netherlands, of the ABN AMRO Bank – one of the leading Dutch banks – is significant in this respect. About four years ago, the bank set up a department for entrepreneurs. Its director, in Amsterdam, provides an interface between local branches and customers, since local bankers are not always able to grasp the needs of people who create their own businesses. Twenty to 25 per cent of the latter are women, and if the loans they receive are smaller (by almost two-thirds) than those granted to men, the bank's explanation for this is that women-owned start-ups tend to be part-time businesses that necessarily have lower turnover and funding requirements.

The Bank of Ireland has a fund of some ECU 60 million for start-up loans to new businesses that fall to qualify for conventional bank lending. It also plays an active role with the "First Step" programme, which provides financing and counselling for small businesses in the start-up phase.

A small percentage of entrepreneurs obtain bank loans at the outset – and there are just as many women as men among them

As a rule, in developed countries, it would seem that just as many women as men are able to obtain bank loans. A 1995 panel study of 30 000 French entrepreneurs by the *Observatoire des Entreprises* showed that 23 per cent of women entrepreneurs had received bank loans *vs.* 22 per cent of men.

Upon analysis, it would appear that the perceived disparity and lack of credibility cited by a large number of women entrepreneurs stem more from banks' wariness of certain lines of business, a certain size of enterprise, and a lack of experience in assessing initial needs and financial planning.

In any event, what these figures show is that 75 per cent of entrepreneurs – men and women combined – did not have any bank-lent funds for their start-up phase (refused by the banks? underestimation of need? absence of need?). In the absence of systematic investigation, the data are insufficient.

Similarly, in Austria, the Institute for Small Business Research, in a study published in 1996, concluded that for 60 per cent of men and women entrepreneurs the greatest problem they faced was access to financing.

Two studies, carried out in Sweden and Switzerland,[6] showed that differences between men and women entrepreneurs emerge primarily when comparisons between the sexes make no allowance for the sectors of activity involved. When female and male entrepreneurs operating in the same sector are compared, the gaps tend to narrow: a man and a woman in the same line of business share more characteristics than two women working in different sectors.

The real issue involved in financing businesses created by women is therefore the ability to borrow small amounts of money. As shown by a study carried out in Germany,[7] although women set up businesses in sectors that require little start-up capital, they encounter greater difficulties in obtaining loans: most lines of credit and aids to SMEs are available only in excess of a minimum capital requirement. One of the problems is that lending small sums of money is proportionately riskier and more time-consuming than making larger loans.

For the past 15 years, schemes running parallel to the banking system have been trying to provide solutions. In addition, the Mama Cash Foundation and Women's World Banking, in order to expand their support capacity, have forged special ties with banks – either directly, through formal agreements, or insofar as their acknowledged planning expertise lends security and reliability to projects that are submitted for financing. To date, these mechanisms have not had any significant impact in the developed economies. To their credit, however, they were pioneering innovators, have devised and tested effective methods and have provided financial solutions to the needs of start-ups.

Should these special measures, tailored to female entrepreneurship, be developed further, or should existing links with the banking system be reinforced? I would argue for the latter course of action.

Micro-loan programmes, many of which are joint projects formulated by specialised financial institutions or by banks that are reaching out to this market, have developed methods that have proven their worth – for small, sole proprietorships in particular – through a comprehensive approach that includes training and monitoring, thus helping to reduce administrative costs:

◊ programmes are based on close relationships between lenders and borrowers, which in some cases can even involve the borrower's being asked, or applying, to become a member of the lending institution;

◊ they generally include financial planning support;

◊ they generally propose (or require) subsequent monitoring;

◊ in these ways, they instil a relationship of trust with borrowers;

◊ even modest financial contributions can supply leverage, provide a supplement or, in some cases, constitute the only source of financing available;

◊ they enjoy fairly high pay-back rates.

Personal contributions and access to guarantees

Guarantee mechanisms can sometimes make up for a lack of collateral *vis-à-vis* banks, but they do not alleviate the chronic shortage of equity that is typical of this type of business.

We have noted that women are more reluctant than men to invest family assets in their businesses, and that banks almost automatically ask for guarantees from spouses, personal security or a pledge of real estate. While it is true that a man who marries under the law of community or property is required to obtain his wife's co-signature to pledge the family's assets, society is less accepting of a woman who gives up family assets, whether on her own behalf or on that of the family.

A few years ago, the World Bank claimed that women owned 1 per cent of the property in the world. It would be very interesting to know the proportion of property that is owned outright by women in each of the OECD countries. Not to mention the discriminatory banking provisions that exist in certain countries, unequal rights with regard to inheritance, unequal access to assets in the event of divorce, or the lack of legal status for women who work with entrepreneur husbands and their inability to lay claim to the business.

The problem of guarantees for small businesses remains unresolved.

Furthermore, the vast majority of guarantee funds do not deal with small SMEs, covering, only those loans that exceed certain amounts. Neither do they cover start-ups, and only rarely do they deal with firms in the service or retailing sectors, in which a majority of women entrepreneurs are involved.

Special-purpose guarantee funds open to all small businesses (such as the Loan Guarantee Scheme in the United Kingdom) and funds reserved for women's enterprises [such as the *Fonds de Garantie aux Initiatives Féminines (FGI)* in France, and the scheme set up by Women's World Banking] fall far short of needs, or else the way in which they operate does not meet with the approval of banks.

Such is the case of the FGIF, initially created for women unable to offer any guarantees. The fund's organisation sets it apart from the operation of other funds and is in sharp contrast to normal banking conventions: the fund itself has a low endowment, and the banks whose loans it guarantees do not have the right to ask for additional guarantees from the borrowers.

Policy implications

First-time entrepreneurs – and to an even greater extent women first-time entrepreneurs – are an economic and social phenomenon of our times.

We know that the formidable growth of women-led businesses around the world is in a great part a response to changes in our societies and the heralded decline of dependent employment, and that, in this context, starting one's own business is a dynamic way of creating one's own job.

Clearly the gap is widening between the requirements of this new model of development and the constraints of the traditional banking system.

From this standpoint, access to financing is of strategic importance, and a closer look at the funding of the type of businesses that women tend to create is required. All of the available research indicates that the lack of suitable financing represents a major obstacle. On average, half of all SMEs fail in their first three years of existence due to initial under-capitalisation that is never remedied, excessive costs due to regulation which imposes a structural handicap,[8] or the unsuitability of the financing terms on offer (*e.g.* rates, credit line closures, etc.).

Our societies will have not only to rely on the entrepreneurial capacity of women, with the employment potential they represent, but also encourage it by promoting businesses supplying local markets with neighbourhood services and by fostering innovative activities (products or services) with a potential for growth.

This new phenomenon requires:

◊ *A private sector, financial approach tailored to the needs of micro SMEs.* In particular, first-time entrepreneurs require financial products that combine savings (that earn suitable returns) with credit (at affordable rates of interest).

◊ *Creation by the large nation-wide banks of departments specialising in micro SMEs and advising entrepreneurs* (improving the dialogue between banks and SMEs with detailed explanations of financing methods and the various products available, and promoting transparency).

◊ *Development of programmes and lines of micro-loans.*

◊ *Mechanisms to guide borrowers before and after their loans* – at least for the first two years after start-up – linking the intermediary bodies to the financial sector. While most countries have financed training for prospective entrepreneurs (and especially the unemployed), funding is practically never made available for subsequent monitoring. Yet, the success and survival rates of start-ups depend on it.

◊ *A local, regional approach*. Geographic proximity reduces transaction costs for small loans, helps consolidate ties between lenders and borrowers and facilitates tracking. In addition, local communities have everything to gain in terms of employment as more and more small businesses spring up.

◊ *New forms of government intervention* (with the authorities assuming economic responsibility and playing a developmental role rather than doling out assistance).

◊ *Encouragement of women's capacity to create viable enterprises with a potential for growth*, by:

- extending the range of initial training and qualifications to all technical fields;

- developing the supply of financial and managerial training.

In short, this new phenomenon could forge new ties between government, financial institutions and intermediary bodies.

In conclusion, I would say that by exploring the access to financing of small businesses created or run by women – businesses that face special difficulties over and above the usual structural problems faced by small firms – we are helping to find solutions for all small businesses.

From this standpoint, we can say that women are paving the way.

NOTES

1. Observatoire des entreprises de l'Agence nationale pour la Création d'Entreprise.

2. "Les relations banque-entreprises en Europe", MGN, May 1995, study carried out for the Ministry of Industry.

3. *Fourth Annual Report 1996* , The European Observatory for SMEs.

4. Johnston and Storey, "Male and Female Entrepreneurs and their Business: A Comparative Study", in Sheila Alien and Carole Truman (eds.), *Women in Business – Perspectives on Women Entrepreneurs.*

5. "Etude exploratoire pour la création d'outils financiers en direction de I'entrepreneuriat féminin en Europe", *Studio Entreprendre*, carried out for DGV of the European Commission, Equal Opportunities Unit, 1995.

6. *Fourth Annual Report 1996,* The European Observatory for SMEs.

7. *Fourth Annual Report 1996*, The European Observatory for SMEs.

8. "Small and Medium-sized Enterprises. A Dynamic Source of Employment, Growth and Competitiveness in the European Union", report presented by the European Commission for the Madrid European Council.

Chapitre 12

LE FINANCEMENT DES FEMMES CHEFS D'ENTREPRISE, UNE DIFFICULTÉ RÉELLE, NÉANMOINS CONTROVERSÉE

par
Patricia Mahot
Directrice associée de Studio *Entreprend*re
Paris, France

Aborder le problème du financement des entreprises créées et dirigées par les femmes, revient à reconnaître implicitement qu'il existe une différenciation sexuée dans l'accès au financement des entreprises.

... Malgré les caractéristiques communes qu'elles partagent avec les hommes, l'ensemble des études et des observations montrent qu'il y a des domaines dans lesquels les femmes rencontrent des difficultés particulières : la réunion des fonds propres au démarrage et l'accès au financement, viennent en tête des préoccupations des acteurs économiques en contact avec la création d'entreprises par les femmes.

... En se penchant sur l'accès au financement des petites entreprises créées ou dirigées par les femmes, qui cumulent les difficultés structurelles de cette catégorie d'entreprises et des difficultés spécifiques, nous contribuons à produire des solutions pour l'ensemble des petites entreprises.

Aborder le problème du financement des entreprises créées et dirigées par les femmes, revient à reconnaître implicitement qu'il existe une différenciation sexuée dans l'accès au financement des entreprises. C'est ce que nous allons tenter de comprendre ensemble. Quels constats faisons-nous ?

Dans les secteurs du commerce et des services les femmes partagent avec les hommes les difficultés communes aux petites PME

Dans leur grande majorité, les femmes créent et dirigent de petites entreprises. A titre d'exemple, en France, sur 30 pour cent d'entreprises créées par des femmes, seulement 7 pour cent ont plus de 10 salariés[1]. Elles entreprennent principalement dans les secteurs du commerce et des services.

Plusieurs facteurs contribuent à cette concentration dans ces secteurs :

◊ les femmes sont arrivées plus tardivement que les hommes sur le marché du travail, dans des emplois moins qualifiés. Peu d'entre-elles ont une expérience de direction ou d'encadrement ;

◊ le poids culturel qui pèse sur le choix de leurs formations initiales et de leurs qualifications, dans une palette de métiers plus restreinte et moins technique, font que des pans entiers de l'économie restent encore peu abordés par elles.

Cette caractéristique se retrouve parmi les créatrices ; une étude conjointe de l'ANCE et de l'INSEE, en France, montre que la majorité d'entre-elles occupaient antérieurement des emplois de catégorie "employé" dans le secteur tertiaire, ou étaient inactives. De plus, quand elles ont une expérience professionnelle, celle-ci est en moyenne d'une durée plus courte que celle des hommes.

A ce titre, elles connaissent les difficultés que rencontrent aujourd'hui les petites entreprises de ces secteurs, qui présentent une vulnérabilité particulière liée :

◊ à leur situation financière – ces entreprises présentent le plus souvent un manque de fonds propres au démarrage et une fragilité financière peu propice au recours à l'emprunt ;

◊ à leur sensibilité aux fluctuations du marché – faible résistance aux baisses d'activité liées à la conjoncture économique ;

◊ à la personnalité de leurs dirigeants – souvent seuls et peu préparés à être chef d'entreprise.

Les banques rebutent à financer ce type d'activités, aussi bien en création qu'en développement. Souvent sous-capitalisées, elles offrent peu de visibilité ni de perspectives de développement, et ce, dans des secteurs où l'essentiel de la valeur de l'entreprise repose sur son dirigeant. L'évaluation des risques pour le banquier se fait souvent comme un pari.

Une étude commandée par le ministère français de l'Industrie en 1995[2], comparative avec le Royaume-Uni, l'Allemagne, l'Espagne et l'Italie, explorait la question suivante : les défaillances d'entreprises ont-elles un lien direct avec le crédit ? Elle faisait apparaître dans ses conclusions que :

◊ il n'existe pas de problème de rareté du crédit, mais un problème de distribution vers les petites entreprises ;

◊ les garanties coûteuses et le coût de gestion de ces entreprises alourdissent les taux auxquels elles empruntent ;

◊ les produits bancaires existants sont inadaptés à la demande ;

◊ le système bancaire ne sait pas répondre à la diversité des petites entreprises en Europe, qui constitue une population disparate en termes de taille et de secteurs d'activités ;

◊ le coût de l'échec, enfin, est très lourd pour les collectivités : en moyenne, en Europe, une entreprise qui se crée, crée 1.7 emploi, une entreprise qui meurt détruit 3.2 emplois.

Les caractéristiques du financement des entreprises créées par les femmes

Malgré les caractéristiques communes qu'elles partagent avec les hommes, l'ensemble des études et des observations montrent qu'il y a des domaines dans lesquels les femmes rencontrent des difficultés particulières : la réunion des fonds propres au démarrage et l'accès au financement, viennent en tête des préoccupations des acteurs économiques en contact avec la création d'entreprises par les femmes. Si tel est le cas, pourquoi ?

Précisons tout d'abord, que le monde des femmes chefs d'entreprises n'est pas un monde homogène. Il recouvre au moins deux grands types de population :

◊ Des femmes chefs d'entreprise, qui n'ont pas de véritable métier : créer est pour elles la seule solution, ou qui ont un métier et une pratique professionnelle, qu'elles valorisent dans une entreprise de service aux entreprises. Elles créent plutôt dans du tertiaire souvent précaire.

  La taille de ces petites entreprises est un état et non un stade de développement. Dans ce cas, les femmes créent moins souvent que les hommes en société et plutôt sous statut "en nom personnel".

◊ Des femmes chefs d'entreprise, plutôt issues d'un milieu familial entrepreneurial, pour qui l'entreprise est un choix, ou qui reprennent des affaires familiales. Elles créent plutôt dans le commerce avec un taux de réussite assez élevé.

  Enfin, celles qui développent une entreprise avec des stratégies de conquête. Elles créent des services ou des produits nouveaux à valeur ajoutée. Ces petites entreprises ont un potentiel de développement.

Mais l'ensemble des observations mettent en avant un certain nombre de caractéristiques, partagées majoritairement par les femmes : est-ce une idée reçue ou bien un état de fait qui résulte de la combinaison de facteurs peu explorés de manière systématique ?

◊ Les femmes ont tendance à prendre moins de risques et à solliciter moins de crédits que les hommes

- Des études réalisées aux Pays-Bas, en Allemagne et au Royaume-Uni[3] ont en effet montré, que les femmes démarraient leur activité, avec des prêts dont le montant était inférieur à ceux sollicités par les hommes. Que ce soit en création ou en reprise d'entreprises, le capital mobilisé est également plus faible que celui des hommes (ANCE/INSEE 1996).

  Le fait qu'elles créent des entreprises plus petites, explique en partie le fait qu'elles sollicitent moins de crédit que les hommes. Sans doute, ont-elles aussi tendance à sous-estimer leurs besoins, en comptant sur leurs propres forces pour se développer à pas comptés : sauf la typologie des femmes "développeurs", la plupart des créatrices ambitionnent de durer, sans forcément viser l'expansion de leur entreprise.

- Elles ont moins recours aux services des banques et préfèrent compter sur leurs propres ressources ou faire appel à leur entourage familial et amical (une étude réalisée au Royaume-Uni en 1993[4], montrait que 55 pour cent des hommes avaient eu recours à un prêt bancaire contre 33 pour cent des femmes).

◊ Elles ont des difficultés à présenter des stratégies de développement sécurisantes pour les organismes financiers

- Ayant moins d'expérience professionnelle, moins de pratique de gestion et de direction, créant plus que les hommes dans des métiers pour lesquels leurs qualifications sont moins adaptées, des entreprises plutôt plus petites et sur des marchés de proximité, une majorité de femmes entrepreneurs ne sont pas en mesure de présenter un prévisionnel de leur activité.

- Elles peuvent apparaître d'autant moins rassurantes pour les organismes financiers que l'Observatoire européen des PME, dans son rapport de 1996, fait état d'un taux de survie légèrement inférieur pour les entreprises créées par les femmes, cette situation résultant des secteurs d'activités choisis par elles.

◊ Elles conservent souvent, en plus de la gestion de l'entreprise, le poids de la gestion familiale (manque de temps et faible incitation à la prise de risque).

◊ Sans compter les différences de comportement entre les hommes et les femmes. Ces dernières investissent plutôt la sphère relationnelle, recherchent des solutions plutôt qu'elles ne produisent des modèles de gestion, gèrent leur entreprise de manière pragmatique et "raisonnable" plutôt qu'elles ne conçoivent des projections financières de leur entreprise.

Ceci est amplifié du fait des motivations qui, bien souvent ont conduit à la création d'entreprise : la création de son propre emploi.

Ces décalages hommes/femmes, entre les formations, les métiers et les pratiques de l'encadrement sont présents à l'esprit des banquiers, aussi bien que dans toutes les strates de la société et contribuent à créer des préjugés défavorables.

L'ensemble de ces données tend à établir que la différenciation avec les hommes portent moins sur l'accès au financement en tant que tel - cette difficulté est le fait de l'ensemble des petites PME en général – que sur le fait qu'elles reproduisent dans l'entrepreneuriat les situations antérieures connues et vécues (type d'emploi, qualification professionnelle moins grande, ambition professionnelle et économique moins grande, sous-estimation des besoins financiers au démarrage, etc.).

Les différences existant dans l'entrepreneuriat sont pratiquement les mêmes que celles existant sur le marché du travail. L'identité professionnelle et les comportements culturels qui y sont associés sont des facteurs clefs pour la compréhension de la problématique. Alors que le niveau de formation initiale des femmes chefs d'entreprise est, dans la plupart des pays de l'Union européenne, équivalent ou supérieur à celui des hommes, elles présentent plus souvent que les hommes un manque de compétences dans les domaines financiers et de la gestion.

Les difficultés des femmes face aux trois principaux besoins de financement des petites entreprises

Les fonds propres

L'existence des capitaux propres est la clef de voûte de toute entreprise : ils lui garantissent son indépendance financière vis–à–vis des banques et lui permettent d'emprunter.

Au cours de ces dernières années, l'accroissement du nombre de créations d'entreprises, notamment par les femmes, a souvent été une réponse à la création de son propre emploi : créées dans l'urgence, dans une situation économique précaire, et l'obligation de s'insérer économiquement, les fonds propres de ces entreprises font très souvent défaut.

Les différences entre les niveaux de rémunération dans les emplois occupés précédemment, expliquent aussi, en partie, la plus grande difficulté des femmes à réunir les capitaux propres suffisants au démarrage.

Dans la logique financière, il n'y a pas de crédit sans épargne, comment alors renforcer les fonds propres de ces petites entreprises, sachant qu'ils peuvent être constitués de plusieurs façons, dont je propose ici un bref récapitulatif ?

a) • l'épargne du créateur
• l'apport des associés
• le prêt d'honneur à la personne

b) • la prise de participation en capital (capital-risque de la part d'investisseurs extérieurs du secteur formel ou informel)

les quasi fonds propres

- subvention, prime à la personne ou à l'entreprise (assimilé à des fonds propres)
- prêts participatifs

L'apport de fonds propres par des interventions en capital-risque du secteur formel

Il semble difficile pour des investisseurs extérieurs de s'engager dans des petites PME, des secteurs du commerce ou des services, sans potentiel de développement, du fait de la difficulté qu'il y a à en sortir, objectif visé par les sociétés de capital-risque. D'ailleurs l'expérience montre que ce mode de financement n'a pas d'incidence sur les petites entreprises, les montants moyens investis par les "capital-risqueurs" étant bien trop importants par rapport aux besoins d'une petite entreprise.

L'apport de fonds propres par des interventions de la part d'investisseurs informels

L'exemple des *business angels*, plutôt répandu dans les pays de culture anglo-saxonne est fondé sur des modes d'intervention plus adaptés, encore que, si les investisseurs ne sont pas purement guidés par une logique de retour sur investissement, mais d'aide au développement et à la pérennité des petites entreprises, en s'engageant d'ailleurs directement dans la gestion de celles-ci, une grande partie des activités économiques dans lesquelles les femmes entreprennent, ne se prêtent guère à la prise de participation en capital.

Sans oublier que dans leur majorité, les créateurs souhaitent conserver leur indépendance et ont peu de volonté d'ouverture de leur capital à des personnes étrangères.

Depuis une quinzaine d'années, des outils privés, souvent pionniers, ont tenté d'apporter des réponses à ce premier besoin de financement : les CLEFES en France, Goldrausch en Allemagne, pour ne citer qu'eux. Les CLEFES offrent un cadre juridique privé pour la constitution d'une épargne de proximité dont les fonds recueillis sont prêtés à une créatrice choisie par le groupe d'épargnants. Cette formule, pour se développer, aurait besoin d'être renforcée par un partenariat bancaire local. Goldrausch, pour sa part, délivre des prêts sans intérêts à des créatrices dont le projet ne satisfait pas aux critères bancaires classiques.

Certains États ont mis en place des programmes d'aide pour aider les femmes à se constituer un capital de départ. Pour ne prendre que le cadre européen, le programme ILE (Initiative Locale d'Emploi), non reconduit, est intervenu, à partir de 1986-87, sous forme de prime au développement d'un projet d'entreprise innovant ou à la création d'emplois féminins, sur des projets sélectionnés. Cette prime directe de la Commission Européenne aux créatrices pouvait être apparentée à un apport de quasi fonds propres.

Le problème des fonds propres reste difficile à résoudre. Force est de constater que les formules mises en place ont un impact très limité[5].

Le financement par l'emprunt

Le crédit reste la principale source de financement de la petite entreprise.

Or, cette clientèle n'intéresse pas vraiment les banques, le faible rapport coût sur rentabilité directe, le risque pris, notamment celui de la création, représentent autant de freins. Les coûts transactionnels ne sont pas proportionnels à la somme prêtée, ils pèsent même davantage pour une petite somme que pour un montant important. La conséquence est que les petits emprunteurs se voient appliqués des taux d'intérêt beaucoup plus élevés.

Aujourd'hui, on constate que les grandes entreprises ont leur propre marché, les moyennes investissent peu et sollicitent donc moins de crédits, la petite entreprise devient une clientèle nécessaire pour les banques.

Bien sûr, le mouvement est timide, souvent assorti de conditions (la proximité par exemple), et, dans la plupart des cas, plutôt le fait des grands réseaux que des petites banques qui elles, au contraire, accentuent leur sélection.

Dans les faits, cela se traduit par un effort vers une approche segmentée de clientèle, dont l'objectif est de mieux adapter les services offerts par les banques à leur clientèle d'entreprise.

On retrouve cette préoccupation dans la plupart des pays européens.

L'exemple, des Pays-Bas, et de l'ABN-AMRO Bank, l'une des banques hollandaises les plus importantes, est à cet égard significatif. Cette banque a ouvert il y a environ quatre ans un département pour les créateurs d'entreprises. Son responsable, à Amsterdam, joue également un rôle d'interface entre les agences locales et les clients (celles-ci n'étant pas toujours à même d'appréhender les besoins des créateurs). Parmi eux, 20 à 25 pour cent sont des femmes et si le montant des prêts qui leur sont accordés est inférieur à celui des hommes (presque un rapport de 1 à 3), les raisons invoquées par la banque sont que les femmes créent le plus souvent à mi-temps, divisant ainsi leur chiffre d'affaires et leurs besoins.

La Banque d'Irlande dispose d'un fonds d'environ 60 millions d'écus pour des prêts affectés au financement de créations d'entreprises qui ne satisfont pas aux critères bancaires classiques. Elle est aussi partie prenante d'un programme avec First Step, pour associer financement et conseil à des petites entreprises en phase de démarrage.

Un petit pourcentage de chefs d'entreprise obtiennent un prêt bancaire à la création, les femmes et les hommes y sont aussi nombreux.

D'une façon générale, dans les pays développés les femmes semblent être aussi nombreuses que les hommes à obtenir un prêt bancaire. L'étude réalisée en France par l'Observatoire des Entreprises en 1995 (panel 30 000 créateurs) montre que 23 pour cent de femmes créatrices ont bénéficié d'un prêt bancaire, pour 22 pour cent d'hommes.

A l'analyse, le sentiment de différenciation ou de manque de crédibilité dont de nombreuses femmes chefs d'entreprise font état, semble plus résulter d'une méfiance à l'égard de certains secteurs d'activité, d'une taille d'entreprise et du manque d'expérience dans l'évaluation des besoins au démarrage et dans leur formulation financière.

Ce que nous enseigne en tous cas ces données, c'est que 75 pour cent des créateurs, hommes et femmes confondus, n'ont pas bénéficié de prêts bancaires au moment de la création (refus de la part des banques, sous-estimation de leurs besoins ou absence de besoins ?) Faute d'une exploration systématique, les données manquent.

Dans le même ordre d'idée, en Autriche, l'Institut de Recherche pour les Petites Entreprises, dans une étude parue en 1996, concluait que pour 60 pour cent des hommes et des femmes chefs d'entreprise, le problème le plus important était bien celui de l'accès au financement.

Deux études conduites en Suède et en Suisse[6] montrent que les différences entre hommes et femmes entrepreneurs s'établissent essentiellement lorsque la comparaison faite entre les sexes ne tient pas compte des secteurs d'activité. Lorsque l'on compare les femmes et les hommes entrepreneurs au sein d'un même secteur, les écarts tendent à se réduire : un homme et une femme, dans le même secteur d'activité, sont plus semblables que des femmes travaillant dans des secteurs différents.

Traiter de la problématique du financement des entreprises créées par les femmes revient donc bien à traiter de l'accès à des petits montants de prêts. Comme le montre une étude réalisée en Allemagne[7], bien que les femmes entreprennent dans des secteurs faiblement demandeurs de fonds au démarrage, elles rencontrent, *a contrario*, plus de difficultés à obtenir un prêt : la plupart des lignes de crédit ou des mesures en faveur des PME ne sont accessibles qu'à partir d'un minimum de capital requis.

Or, la distribution de petits prêts est proportionnellement plus coûteuse en temps consacré et en risques qu'elle ne l'est pour des montants plus importants.

Depuis une quinzaine d'années, des dispositifs parallèles au système bancaire ont tenté d'apporter des réponses. La Mama Cash Foundation et la Women's World Banking ont d'ailleurs, pour élargir leur capacité d'intervention, noué des relations privilégiées avec les banques, soit en ayant passé un accord avec elles, soit parce que leur expertise en amont, est reconnue et apporte caution et fiabilité au projet proposé pour un financement. Jusqu'à présent, ces outils n'ont pu étendre leur impact dans les économies développées. Mais, ils ont le mérite de s'être engagés très tôt sur les voies de l'innovation, d'avoir expérimenté et construit des méthodologies efficaces et proposé des réponses financières aux besoins des nouvelles entreprises.

Faut-il développer ces mesures particulières, dédiées à l'entrepreneuriat féminin ou faire progresser les rapprochements avec le système bancaire ? Je plaiderais en faveur de la seconde voie.

Les programmes de micro-prêt, souvent mixtes, développés, soit par des organismes financiers spécifiques, soit par des banques qui s'ouvrent à ce marché, ont mis au point des méthodes qui ont fait leur preuve, notamment pour des petites entreprises individuelles, par une approche globale, intégrant formation et suivi, favorisant ainsi la réduction de leurs coûts transactionnels :

◊ ils reposent sur une relation prêteur/emprunteur étroite, qui peut même parfois se fonder sur une cooptation, ou une adhésion à l'organisme prêteur, en tant que membre ;

◊ ils incluent le plus souvent un accompagnement au montage du projet d'entreprise ;

◊ ils développent le plus souvent (ou rendent obligatoire) un suivi des dossiers après intervention financière ;

◊ ils instaurent par ces différents moyens une relation de confiance avec l'emprunteuse ;

◊ même modestes, ces apports financiers peuvent faire effet de levier, être complémentaires ou, parfois, représentent la seule source de financement accessible ;

◊ ils obtiennent des taux de remboursement assez élevés.

L'apport et l'accès aux garanties

Les dispositifs de garantie peuvent parfois suppléer l'absence de cautions vis–à–vis des banques, mais ils ne résolvent pas les carences chroniques en fonds propres de ce type d'entreprise.

Nous constatons que les femmes répugnent plus que les hommes à engager le patrimoine familial dans leur activité professionnelle et que les organismes bancaires sollicitent quasi-automatiquement, les garanties de leur conjoint, cautions personnelles ou garanties immobilières. Il est vrai que dans le cas du mariage sous le régime de la communauté de biens, un homme doit nécessairement obtenir la co-signature de sa femme, pour engager le patrimoine familial. Mais il reste qu'il est culturellement moins bien accepté pour une femme d'aliéner le patrimoine familial, que ce soit de son fait à elle ou du fait de la famille.

Il y a quelques années, la Banque Mondiale affirmait que les femmes détenaient 1 pour cent de la propriété mondiale. Il serait très intéressant de connaître la part de propriété en propre, détenue par les femmes dans chacun des pays de l'OCDE.

Sans compter les mesures bancaires discriminatoires qui existent dans certains pays, l'inégalité des droits devant l'héritage, l'inégalité d'accès des femmes au patrimoine en cas de divorce, ou encore l'absence de reconnaissance du statut des femmes collaboratrices de leur mari entrepreneur et leur impossibilité à faire valoir leurs droits à l'entreprise.

Le problème de la garantie des petites entreprises reste entier.

D'ailleurs, dans leur grande majorité, les fonds de garantie ne s'adressent pas aux petites PME, n'intervenant qu'au-dessus de certains montants. Ils n'interviennent pas non plus à la création et ne s'adressent que très rarement aux entreprises des secteurs des services et du commerce, secteurs dans lesquels s'engagent une majorité de femmes entrepreneurs.

Des fonds de garantie spécifiques intervenant sur l'ensemble des petites entreprises, comme le Loan Guarantee Scheme au Royaume-Uni, ou dédiés aux entreprises de femmes, comme le Fonds de Garantie aux Initiatives Féminines en France (FGIF), ou le dispositif mis au point par la Women's World Banking, sont loin de couvrir les besoins, ou encore le contexte dans lequel ils ont été conçu, a développé une logique de fonctionnement qui ne rencontre pas la faveur des banques.

C'est le cas du FGIF, créé initialement pour des femmes ne pouvant présenter aucune garantie. Son mode d'organisation le différencie du fonctionnement des autres fonds et le marginalise du point de vue de la logique bancaire classique : le fonds lui-même a été faiblement doté, les banques n'ont pas le droit de solliciter d'autres garanties complémentaires auprès des emprunteuses.

Implications politiques

Les nouveaux entrepreneurs, *a fortiori* les nouvelles entrepreneuses, constituent un phénomène économique et social contemporain.

Nous savons que le formidable accroissement des entreprises de femmes dans le monde répond pour une grande part aux mutations de nos sociétés et à l'annonce du déclin de l'emploi salarié et que, dans ce contexte, la prise d'initiative économique constitue une réponse dynamique à la création de son propre emploi.

Il est clair que le décalage s'amplifie entre les besoins de ce nouveau modèle de développement et les contraintes du système bancaire traditionnel.

L'accès au financement est, de ce point de vue, stratégique. Il conviendrait de se pencher de plus près sur le financement du type d'entreprises que créent les femmes. L'ensemble des études indiquent que l'absence de financement adaptés représente un obstacle majeur. En moyenne, une PME sur deux échoue au cours de ses trois premières années d'existence, du fait d'une sous-capitalisation initiale jamais reconquise, des coûts excessifs dus à la réglementation qui entraîne un handicap structurel[8] et des conditions financières peu adéquates qui lui sont faites (taux, suppression de lignes de crédit, etc.).

Nos sociétés devront compter, non seulement sur la capacité des femmes à entreprendre avec le potentiel d'emplois qu'elles représentent, mais encore la favoriser, aussi bien pour des entreprises agissant sur un marché local, dans des services de proximité, que pour des activités innovantes (produits, services) à potentiel de développement.

Ce nouveau phénomène a besoin :

◊ *d'une approche privée, financière, mais adaptée, révisée pour les besoins spécifiques des petites PME.* Les nouveaux entrepreneurs ont besoin, en particulier, de produits financiers qui lient l'épargne (rémunérée à des taux convenables) et le crédit (à des taux supportables) ;

◊ *de la création dans les grands réseaux bancaires de départements spécialisés dans les petites PME et l'accueil aux créateurs* (améliorer le dialogue banques/petites PME par des explications précises sur les différents produits disponibles et sur les modes de financement, favoriser la transparence) ;

◊ *du développement des programmes et des lignes de micro-prêts* ;

◊ *de mécanismes d'encadrement en amont et en aval du crédit*, les deux premières années de création au moins, en reliant les organismes intermédiaires au secteur financier. Si la plupart des États ont financé la formation en amont de la création (surtout pour les chômeuses-créatrices), le suivi des entreprises n'est pratiquement jamais financé. Le taux de succès et de pérennité des entreprises en dépend pourtant ;

◊ *d'une approche locale, régionale* : la proximité géographique est facteur de réduction des coûts de transaction pour des prêts de faible montant, de consolidation du lien prêteur/emprunteur, de suivi des entreprises. De plus, le tissu local a tout à gagner en termes d'emplois, au fur et à mesure que s'enrichit le tissu des petites entreprises ;

◊ *de renouveler les formes de l'intervention publique* (responsabilité économique des pouvoirs publics, assurant un rôle de développeur et non d'assistance) ;

◊ *de favoriser la capacité des femmes à créer des entreprises viables et à potentiel de développement*, par :

- l'élargissement de la palette des formations initiales et de leurs qualifications à l'ensemble des domaines techniques ;

- le développement des offres de formation financières et de gestion.

En bref, ce phénomène nouveau pourrait déboucher sur un lien pouvoirs publics, organismes financiers, organismes intermédiaires.

Pour conclure, je dirais qu'en se penchant sur l'accès au financement des petites entreprises créées ou dirigées par les femmes, qui cumulent les difficultés structurelles de cette catégorie d'entreprises et des difficultés spécifiques, nous contribuons à produire des solutions pour l'ensemble des petites entreprises.

De ce point de vue, nous pouvons dire que les femmes ouvrent des voies.

NOTES

1. Observatoire des entreprises de l'Agence nationale pour la Création d'Entreprise.

2. "Les relations banque-entreprises en Europe", société MGN, mai 1995, réalisée pour le compte du ministère de l'Industrie.

3. *Fourth Annual Report 1996,* The European Observatory for SMEs.

4. Johnston et Storey, "Male and Female Entrepreneurs and their Business: A Comparative Study", dans Sheila Allen et Carole Truman (éds.), *"Women in Business - Perspectives on Women Entrepreneurs*".

5. Etude exploratoire pour la création d'outils financiers en direction de l'entrepreneuriat féminin en Europe", Studio *Entreprendre*, pour le compte de la DGV de la Commission européenne, Unité pour l'égalité des chances, 1995.

6. *Fourth Annual Report 1996,* The European Observatory for SMEs.

7. *Fourth Annual Report 1996,* The European Observatory for SMEs.

8. "Le rôle des petites et moyennes entreprises comme source dynamique d'emploi, de croissance et de compétitivité dans l'Union européenne", rapport présenté par la Commission européenne au Conseil européen de Madrid.

Chapter 13

SMEs IN THE PHILIPPINES

by
Zenaida Gonzales Gordon
President, Gordon Enterprise,
Philippines

There is little doubt that women are making a difference in the South-East Asian economies. They make up between one-third to almost one-half of the workforce in five countries, namely the Philippines, Indonesia, Malaysia, Thailand and Singapore, where the highest labour participation rates are to be found. Women play a particularly significant role in the enterprise sector where rapid economic growth and industrialisation have increased employment opportunities both in industry and in the services. However, the most vigorous growth has been in entrepreneurship. More and more women are becoming entrepreneurs out of necessity in order to augment their family's income. Some of these self-employment activities are an expansion of women's domestic skills, such as cooking, sewing, cleaning, and so on. Often these women start their businesses as a single individual working for her own account out of her own home.

Many favourable factors have facilitated the participation and contribution of women in the growth of Philippine enterprises. These include their access to education, the increasing ease of travel and communications, the advances and adoption of technological change and innovations, and the positive impact of the rise in per capita incomes. The entry of women into business has also become increasingly acceptable because of the undeniable economic benefits of the two-income family. The supportiveness of the family also plays an important role in the effectiveness of Philippine women in enterprises. The extended family system provides support that enables women to succeed in their undertaking while still fulfilling their domestic obligations. The strength of Asian family ties is a major factor in the success of women entrepreneurs.

The significant role played by SMEs as critical engines for growth is well recognised in the Asia-Pacific region, in particular their ability "to respond flexibly to technological advances and the diversification of consumer needs". In the Philippines, the SME sector is crucial to the overall growth and development of the economy. SMEs make up over 90 per cent of manufacturing establishments and account for over 50 per cent of the country's employment. As of the last census, there were 224 194 establishments operating in the Philippines, in both the manufacturing and service sectors (as shown below).

Figure 1. Distribution of SMEs

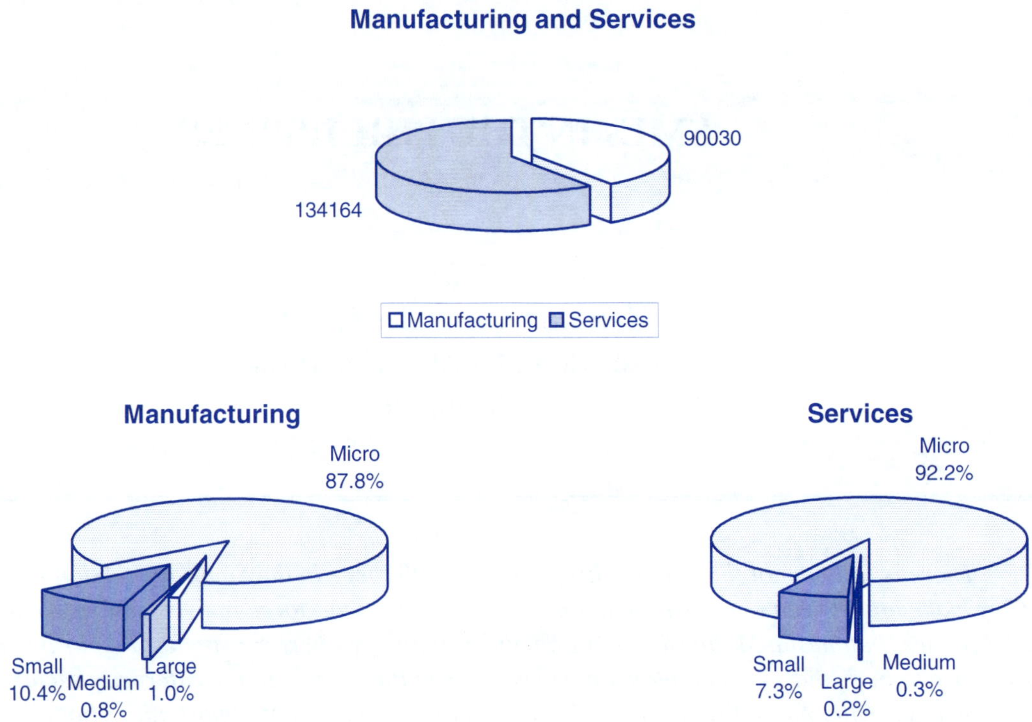

Source: DTI table from National Statistical Office data.

Of these, 1 152 were large ventures, and 1 114 were in the medium-sized category. The balance consisted of micro, cottage and small enterprises or those with capital assets of less than 15 million pesos (slightly more than US$0.5 million), which are generically classified under the "small" business category. The ratio more or less holds true when only manufacturing SMEs are considered. The above data show a preponderance of micro and small businesses in manufacturing – a total of 88 382 firms *vs.* 895 large and 753 medium-sized businesses.

While SMEs clearly outnumber the country's large companies by a ratio of nine to one, it is the larger firms that account for the bulk of sales turnover. Total revenues of all business establishments, including both manufacturing and services, amounted to P 934 billion in 1991. Of this total, the manufacturing sector accounted for 88 per cent (or P 822.2 billion), and the large firms from this sector earned P 590.7 billion (72 per cent of the total) *vs.* P 231.6 billion generated by SMEs, including micro ventures. This data also shows the distribution of SMEs in manufacturing, with 40 465 (45.40 per cent) firms in food, and 14 135 (15.86 per cent) in wearing apparel, out of a total of 89 135 establishments.

Figure 2. SME revenues, in billion pesos, 1993

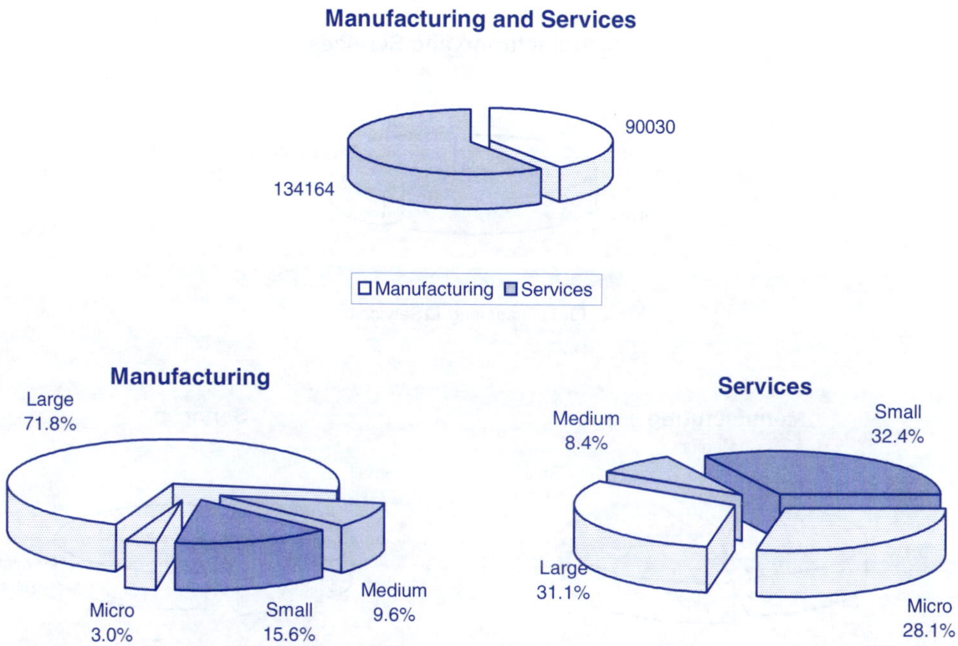

Source: Adapted from NSO data.

Table 2. Distribution of SMEs in manufacturing

By sector, 1993

Pottery, china and earthernwear	456	Footwear	1 709
Textiles	1 894	Machinery, excluding electrical	2 090
Wood and cork products	2 376	Other non-metallic mineral products	2 812
Furniture	5 719	Fabricated metal products	8 714
Wearing apparel	14 135	Food	40 465
Others	8 811		

However, the reverse is true for employment generation, because here it is the SMEs, including the micro ventures, that create the largest number of jobs – and not the large firms that accounted for the largest revenues. SMEs generated 1 438 643 jobs *vs.* 666 501 for the largest enterprises. This indicates that SMEs contributed almost 55 per cent of the total number of jobs generated by all types of business establishments. Again, this percentage remains consistent when only manufacturing establishments are analysed. The data also show that micro ventures create the largest number of jobs. By industry sector, food processing generated the most jobs among manufacturing SMEs, followed by the wearing apparel, fabricated metal products and furniture. The food sector contributed 18.6 per cent of the total number of jobs generated by the manufacturing sector in 1991, with wearing apparel standing at 8 per cent.

Figure 3. Employment generation: SMEs in services and manufacturing, 1993

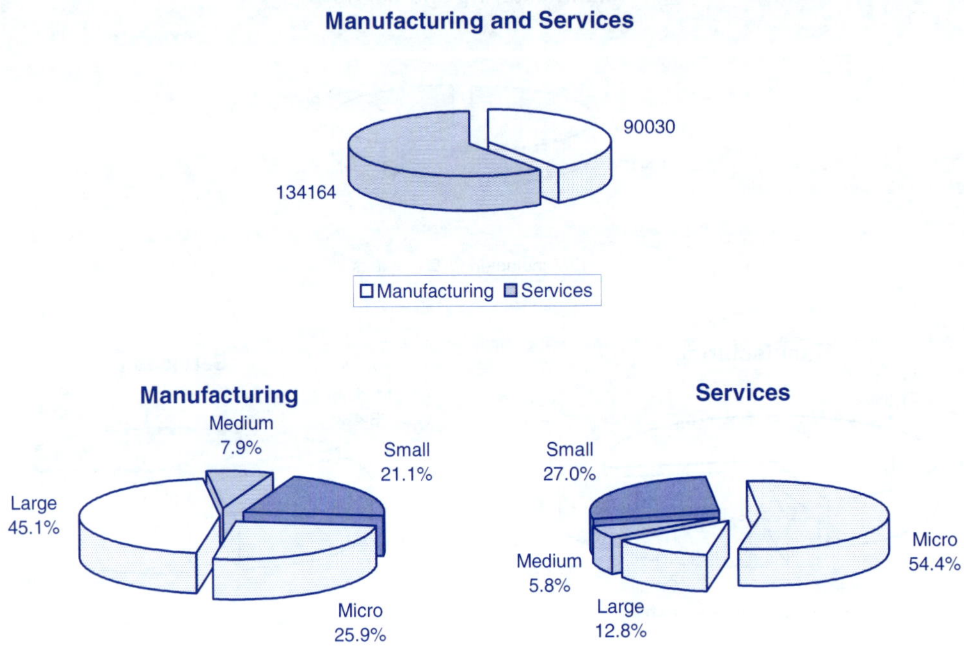

Source: Adapted from NSO data.

All regions of the Philippines showed increases in the number of SME locators. Of the total number of existing SMEs in the manufacturing sector, some 80 per cent are located outside of Metro Manila. The regions benefitting from more developed infrastructure in terms of power, utilities, transport facilities and ready sites where small investors can quickly set up factory operations, were generally more successful in attracting SME start-ups.

Since SMEs are more likely to be labour-intensive, they are critical to dispersing new industries across the countryside, thus bringing about a more balanced agro-industrial growth and greater equity in income distribution.

Over the past five years, the Philippine Government under President Fidel V. Ramos has intensified the policy climate and programme support mechanisms for SMEs. Basic legislation was passed through the Republic Act 6977 – the Magna Carta for SMEs. This law established two core institutions – the SMED Council to advise government on SME policy; and the Small Business Guarantee Finance Corporation, which was created to give small firms access to low-cost credit sources. On the global front, the Philippine Government has also committed itself to building a better operating environment for SMEs, as evidenced by its hosting of the APEC Centre for Technology Exchange and Training (ACTETSME) in Los Banos, which was inaugurated in September 1996. This Centre is expected to become an invaluable information and technology resource – not only for SMEs in the Philippines, but also for all SMEs in the Asia-Pacific region. Indeed, it can be concluded that the stage is being set for Philippine SMEs to be able to survive and look forward to many exciting avenues for growth. It is now up to the Philippine SMEs themselves to rise to the challenge and increase their ability to compete in order to expand their markets in the arena of their choice – domestic, exports, or both.

Women in enterprises

There is little doubt that women are making a difference in the South-East Asian economies. They make up between one-third to almost one-half of the workforce in five countries, namely the Philippines, Indonesia, Malaysia, Thailand and Singapore, where the highest labour participation rates are to be found. Women play a particularly significant role in the enterprise sector where rapid economic growth and industrialisation have increased employment opportunities both in industry and in the services. However, the most vigorous growth has been in entrepreneurship. More and more women are becoming entrepreneurs out of necessity in order to augment their family's income. Some of these self-employment activities are an expansion of women's domestic skills, such as cooking, sewing, cleaning, and so on. Often these women start their businesses as a single individual working for her own account out of her own home.

In the Philippines, women comprise over half, and sometimes as much as 70 per cent, of all those working for their own account in manufacturing, trade and social/community and personal services. A common pattern for many woman entrepreneurs is to expand a craft or business which originally started as a hobby. A woman normally learns to sew her children's clothes and household items in order to save money while the children are growing up. Then, once they have grown up, she graduates to sewing other items for her relatives, friends and neighbours. At a later stagde, she takes advantage of market opportunities, buys two or three sewing machines and hires seamstresses to assist her in her activity. From these humble beginnings, she can move steadily upwards to a fully fledged company employing 50 workers or more if she is willing to take risks, to innovate and develop new products and markets and to participate in new growth areas for so long as market opportunities exist.

Services associated with home-making skills, such as restaurants and hotels, are areas in which Philippine women excel. Trading is also an important area, with some women going beyond trading to succeed in related businesses such as owning and running convenience stores, department stores, supermarkets and similar establishments. Another area where women have made their mark is in real estate marketing and development. Women have also gained ground in the emerging areas of growth and in areas not traditionally associated with home activities or "women's work". Many women are successful in the exciting world of technology, in consumer electronics and durables, in semiconductors, in computers and their applications, in car manufacturing, machinery and equipment and products that protect or conserve the environment.

Many favourable factors have facilitated the participation and contribution of women in the growth of Philippine enterprises. These include their access to education, the increasing ease of travel and communications, the advances and adoption of technological change and innovations, and the positive impact of the rise in per capita incomes. The entry of women into business has also become increasingly acceptable because of the undeniable economic benefits of the two-income family. The supportiveness of the family also plays an important role in the effectiveness of Philippine women in enterprises. The extended family system provides support that enables women to succeed in their undertaking while still fulfilling their domestic obligations. The strength of Asian family ties is a major factor in the success of women entrepreneurs. Many successful businesses in Asia, and in the Philippines for that matter, are family businesses or continue to have major links with specific families. There are no longer barriers to women taking over and/or running a major division of the family business. In today's Philippine business setting, both male and female children – in fact, whoever is interested and capable – can play a major role in the business – that role being limited only by the ability of the individual concerned. Finally, the most important factor contributing to the rise of women in enterprises is the greater self-confidence and empowerment of women themselves!

A special role for SMEs and for women in SMEs

The role of SMEs in the Philippines is not limited to the creation of wealth or capital. SMEs, in addition to stimulating gainful employment, tend to be far more innovative in developing indigenous or appropriate technology which research and development can then turn into pioneering technological breakthroughs.

SMEs are notably skilful in maximising the use of scarce capital resources and are able to partner large firms by supplying locally available raw materials in unprocessed or semi-processed form. In this manner, SMEs effectively increase the local content of the final goods which are processed and marketed by large manufacturing operations. Because they are unhampered by size and the complications of adapting assembly-line equipment to new processes, they are quick to assimilate new design trends, develop contemporary products and bring them to the market-place ahead of the competition. A strong and vigorous SME sector is therefore vital to moving forward the Philippine economic development agenda, particularly in towns and *barangays* where SMEs can inject activity into rural areas and thus become the necessary agents of change in raising income equity among the population.

The potential of women entrepreneurs in SMEs is especially great because women are the ones who look for other sources of family income and are able to easily explore self-employment activities. Filipino women are well-known for their innate artistic skills, sense of balance and colour, and their innovativeness in transforming indigenous materials into products marketable on local as well as international markets. Another notable aspect of the Filipino woman is her ability to attend to her family duties and be sensitive to the needs of the family despite the heavy schedule she has to follow in order to nurture and keep her business going.

Success stories of women in SMEs

Association of Negros Producers (ANP)

The entrepreneurial traits of Filipino women are well demonstrated by the members of the *Association of Negros Producers*, a non-government organisation composed of 75 women and 22 men. They manufacture a range of goods including garments, bags, shoes, travel accessories, gift items, toys and novelties, home decor furnishings, foods and costume jewellery. This association was started by a pioneering group of 47 people (mostly women), who initiated small backyard ventures, mainly to provide jobs to displaced workers (some 150 000 strong) during the sugar crisis in the southern island of Negros. Using indigenous materials such as corn cobs, coco husks and twigs, they produced gift items, novelties and decorative products which were first sold to neighbours, friends and relatives. Support then came from then Governor Daniel Lacson of Negros, and the Department of Trade and Industry agencies – notably the Philippine Trade Training Centre (PTTC), the Product Design and Development Centre of the Philippines (PDDCP) and the Centre for International Trade Expositions and Missions (CITEM). These agencies lent a helping hand by providing export training, product and raw material research and development, quality control and packaging and, most importantly, securing domestic and then foreign markets. A non-revolving credit line of P 3 million from the Technology and Livelihood Resource Centre enabled the association's cash-strapped members to expand their productive capacity.

From its modest beginnings, ANP has nurtured some 25 exporters and has grown to be a leading SME organisation in the country – with women leading, controlling and owning most of the businesses. The output of its members has found its way into the country's large department stores and boutiques, and are regularly shiped to the United States, Germany, France, Japan and other Asian countries. Most important of all, some 12 000 new jobs have been created as a result of the association's activities. The following table presents a percentage breakdown of its producers according to scale (assets in Philippine pesos):

a. micro	< P 150 000	8 members, or 8.25%
b. cottage	P 151 000 – 1.5 million	75 members, or 77.32%
c. small	P 1.5 million – 15 million	10 members, or 10.31%
d. medium	P 15 million – 60 million	4 members, or 4.12%

The problems most commonly encountered by the association members include insufficient credit and difficulty in securing bank loans, lack of good raw material at reasonable costs, and lack of technical information in order to improve production capabilities and thereby achieve economies of scale.

Despite these difficulties, the association has some 22 women members who feel that, given adequate assistance and support, they will be able to cope with the requirements of the market-place. In fact, the following members are already exporting their production in different categories:

1. A&T Garments, smocked children's dresses – Alexandra Y. Francisco

2. Anaware Ceramics, Pandan crochet and ceramics – Anita F. Feria/ Mary Ann Colmenares

C. Bacolod Shellcraft and Artwares, shellcraft, flower arrangements, frames, wedding souvenirs and novelty items – Emily Ann Longa

4. Cross Stitch Co., table and kitchen linen, baby linen, Christmas decorations and gift items – Gina de Leon

5. Christine Sicangco Lighting Design, Inc., lighting fixtures (table lampshades and pendants made of handmade paper) – Christine Sicangco/Ma. Mae Lagaras/Florence Bataan

6. Far East Bamboo Enterprises, bamboo furniture and fixtures, lampshades and home decor – Mr. and Mrs. Emil Lamata

7. Gala Gifts Gaff, Inc., travel and home organisers – Paloma V. Isas/Letty A. Corriente Las Palomas

8. GerardPhilippines, Inc., fashion accessories, boxes, bags – Marites Gonzales

9. Green Harvest Food Industries, fish preparations, coconut gel, sweet pickles – Doris L. Loarca and Isabel M. Patron

10. Hacienda Crafts Co. Inc., serving trays, placemats, baskets and picture frames – Christina B. Gaston

11. Handumanan Workshop, corn husk and cob products (vegetables, flowers, baskets and trays) – Mitos Cooper and Jimmy Gonzales

12. In Hand Negros, Inc., hand-woven placemats, napkins, window shades, dividers, etc. – Anna Anaya

13. Johanna' s Export, furniture, fixtures, gifts and home accessories – Johanna Montilla Lacson

14. JZR Industries, ceramic canisters, dolls, hand-painted containers, jars and vases – Julieta Z. Rodriguez

I5. Kasipagan Association of Mothers' Clubs, Christmas decorations, novelty items and table linen – June M. Villarante

16. Kindlecraft, candles and ceramics – Christine and Joaquin Heras

17. Milflores Finecrafts Corporation, laminated coco shells, sea shells, banana picture frames, jewellery boxes and table tops – Cynthia Flores

con't d

18. Obra Negrense, Inc., candle-holders, planters, kitchen trays, Christmas decorations – Ma. Vicenta P. Rio, Edna G. Yap and Jennifer Gochangco

19. Reeds & Weeds, Pandan baskets, hotmats, placemats, trays, coasters, canisters with ceramic handles and beads – Sylvia Campos

20. Tumandok Crafts Industries, hand-crafted/laminated novelty gifts and houseware – Josephine Locsin/Severino Tambanillo, Sr.

21. Vicmik Enterprises, Inc., children's hand-smocked garments – Isabel V. Lovina

22. VS Heirlooms Pacifica, Inc., hand-crafted/crocheted murals, Christmas decorations, lace on baskets, costume dolls – Suzette and Tisha Agustin

In keeping with the association's mandate of "helping its members to help themselves", it has taken the following strategic steps:

1. It has built its own showroom in the heart of Bacolod City in order to properly exhibit its members' products and facilitate its movement.

2. It has established the ANP Foundation for Enterprise Development to institutionalise its entrepreneurship and technical assistance programmes and bring them to other depressed areas in the country.

3. It has established a policy of sharing experiences and resources, including raw materials, so that large volume orders can be met by pooling production capacity.

4. It carries out training programmes to keep its members up to date on export skills and market trends.

There is no doubt from the above that the women in SMEs in Negros are making a significant contribution to building prosperity for the community. The next challenge is how they will master their resources in order to positively react to the impending threat of globalisation. Special incentives and assistance schemes for SMEs are now available, as are a number of special financing programmes. The stage is therefore set for these women to take on this new challenge. We all expect the ANP women to make the grade, given their track record and the necessary support from their public and private sector allies.

The Department of Trade and Industry's Golden Shell Awardees

Another noteworthy indication of the Filipino women's successful accomplishments in SMEs are the winners of the Golden Shell Awards for "excellence in exports". Over a span of 13 years, a total of 116 winners have been chosen, with some 59 companies (50.8 per cent) having women at the helm or as active partners in the pursuit of success. The winners are selected based on rigid criteria that include design excellence, production efficiency, total quality management and strategic marketing practices. The majority of the awardees are SMEs who bested large and more highly endowed contenders. By hurdling the barriers in competitive world markets, these SMEs have achieved a position significantly different from other Philippine exporters. Some winners have even successfully jumped from the ranks of SMEs to the world of big-time players.

The following are some of the winners of the Golden Shell Awards over the past 13 years in different product categories. The majority are SMEs – owned, led, managed or co-managed by women:

1. Cojac International, 1983	Atty. Cora Cruz. Jacob
2. Clarisse Garment Venture, 1983	Clara Cabarro
3. Lahi Crafts, Inc., 1983	Resty and Lou Lerma
4. Narda's Handwoven Arts & Crafts, 1983	Leonarda Capuyan
5. Shelmed Cottage Treasures, 1983	Sheldon and Mediatrix Villanueva
6. South Sea Designs, Inc., 1983	Mr. and Mrs. Vicente Ang Cho Kok
7. Auralyn's Enterprises, 1984-85	Nely Guan
8. Blue Bell Commodities, 1984-85	Teresita Hao Chin
9. Creative Lines, Inc., 1984-85	Victoria Fabregas
10. FF. International Mfg. Corp., 1984-85	Josie Cruz Natori
11. Mehitabel Furniture Company, 1984-85	Josephine Booth
12. RC Development Corporation, 1984-85	Carmen Yulo
13. Marigold Commodities Corporation, 1987	Clara Reyes Lapuz
14. M.R. Knitwear Specialists Phils., Inc., 1987	Margarita J. dela Rama
15. R & M Preserves, 1987	Elizabeth T. Siao
16. Trimica, Inc., 1987	Lydia Capistrano
17. Victoria Export, Inc., 1987	Manuel and Edna Sta. Cruz
18. A Touch of Green, 1988	Maritess Pineda with Angel Alvendia
19. JB Woodcraft, 1988	Myrna Bituin
20. Jover's Philippines, 1988	Hortensia Dytuco with husband
21. Naturecraft Industries Corporation, 1988	Benigno and Nona Ricafort
22. The Ampersand Corporation, 1988	Rosemarie Zamora
23. AA Export and Import Corporation, 1989	Elena S. Lim
24. Interior Crafts of the Islands, Inc., 1989	Betty Cobompue
25. Linens International Inc., 1989	Sabina Koschinger
26. Saranga Ceramic Contracting Inc., 1989	Elenita San Agustin Sandejas
27. Standard Products Exporter, Inc., 1989	Mrs. Chua and husband
28. Bulacan Garden Corporation, 1990	Mr. and Mrs. Victorino S. Bustos
29. GIDC Export Corporation, 1990	Celia Lazaro
30. Mejore Woodworks, Inc., 1990	Nathalie Go Ong
31. Tenthouse Garments, Inc., 1990	Evelyn Cristobal Salvador
32. Andamar Incorporated, 1991	Corazon and Mariza Andal
31. Roundtree Merchandising Corporation, 1991	Betty Cang
34. Teano International Export, 1991	Veronica Teano
35. Silver Swan Manufacturing Co., Inc., 1992	Gloria Sison
36. Country Export (Phils.), Inc., 1992	Marianna UyKimpang
37. Aquarose Industries, 1993	Joyce Nathalie Yang
38. Asia Embroidery, Inc., 1993	Lourdes Ragasa Chan
39. Craft Mill, Inc., 1993	Jean Bautista
40. Janice Ice Minor Export, 1993	Janice Minor
41. Unson, Faustmann & Co., 1993	Mia Faustmann and Ma. Luisa Unson
42. Victoria Art & Craft, 1993	Flora Pantaleon, Emma Pantaleon, Fely Pantaleon and Nini Pantaleon-Ledesma
43. Accent Pieces Exports, Inc., 1994	Carmelita and Rowena Sembrano
44. Fastech Microelectronics, Inc., 1994.	Charito Montemayor
45. Loran Industries, Inc., 1994	Lorna Quisumbing
46. Mil Export Phils., 1994	Milagros Lacson
47. RFM Corporation, 1994	Lourdes Valenton
48. Sarte, Teodoro & Associates, 1994	Emma V. Teodoro
49. GSG Industries, 1995	Gigi Gonzales
50. Johann's Exports, 1995	Johanna Lacson
51. Locsin's International, 1995	Cindelyn Locsin with husband
52. Touch of Craft, Inc., 1995	Jo Ann Carceller
53. United Coconut Chemicals, Inc., 1995	Evelina Petino
54. Alen International Industrial Corporation, 1995	Mr. and Mrs Alfredo Lagman

cont'd

55. Airiver Export Corporation, 1995	Ma. Fe Caridad de Rivera
56. Chrysara/Adorica, Inc., 1995	Alice Reyes Van Doorn
57. Lara's Gifts and Decors, 1995	Lara Maria Villafuerte
58. Metal Forming Corporation, 1995	Lourdes Borromeo
59. Tentay Foods and Sauces, 1995	Velia J. Cruz

It may be gleaned from the above list that women in SMEs excel in handicrafts, garments, furniture, processed foods, linens, ceramics, bags, fashion accessories and similar products. An exception is Sarte, Teodoro and Associates, where Emma V. Teodoro, its President, and her team of trained staff lead her company to become a supplier of software development and conversion services with state-of-the-art technology and long years of systems development experience. Today, the company is moving gradually towards becoming a provider of packaged solutions that can respond to the vital information technology requirements of business organisations. Anticipating the challenge of the year 2000 "date" problem, or what information technology professionals call the "millennium bug", Ms. Teodoro and her associates have developed FIELDEX, a software tool which will speed up the conversion of programmes to become year-2000-compliant. To market FIELDEX, they have forged alliances with IT outlets in Europe and the United States.

Meycauayan Jewellery Industry Association, Inc.

Another industry in the Philippines where SMEs predominate is the jewellery industry, which is centred in Meycauayan, Bulacan, on the island of Luzon, more than an hour north of Metro Manila. Its association, which was organised in August 1985 has 90 members, with 75 small firms (P 1.5 million to P 15 million capitalisation) and 15 medium-sized firms (P 15 million to P 50 million capitalisation. This industry is yet another major example of women's involvement in SME businesses, with 40 firms in the association being owned by women, 24 being managed by them and the balance of 26 having men on top of the organisation, but with women still making things happen in a supporting role as the second lead person.

Another remarkable situation in this association is the fact that a woman has been its President since 1988 in the person of Ms. Cecilia R. Ramos. Ms. Ramos also runs a jewellery operation in close combination with her husband who handles the "nuts and bolts" while she does the marketing administration and assists in the design aspects of the business. In terms of value, it is estimated that this industry's revenues can run into several billions of pesos, but unfortunately no records can substantiate this as most of the players belong to what is commonly known in the Philippines as the "informal" or "underground" sector. Thanks to Ms. Ramos' leadership and the close co-operation of her Board, the association has been able to convince the lead players to surface and bond themselves to become a strong and vigorous group. As a result the members have become more forward-looking, more marketing-oriented and fully conscious of the fact that they have to learn more and sharpen their skills in order to survive and improve their businesses. Recognising that the Philippines is now on a lower rung of the ladder with Thailand, Hong Kong and Singapore as the leading jewellery manufacturers, the association has embarked on the following activities:

1. Study missions were conducted to improve members' skills in silver filigree production, antique/ethnic design and jade lapidary.

2. Seed funding was obtained through the Department of Trade and Industry in co-operation with the provincial government of Bulacan, to establish the first *and only* successfully run jewellery-making centre in the Philippines. To-date, this training centre has graduated 1 600 trainees who are all currently employed in various jewellery shops, not only in Meycauayan but throughout the country and abroad.

3. The Association has participated in various marketing activities and domestic fairs that generated interest and awareness of its members' expertise. In 1990, the Association won the prestigious and coveted KATHA Award for best product design during the 12th Manila Fame Exhibition, for its reproductions of Philippine antique jewellery in gold and silver.

4. The Association made a first attempt to test its ability to penetrate the export market when Ricel's Jewellery (owned by Ms. Ramos) joined the Fine Jewellery Selling Mission, first to Osaka in September 1990 and September 1991, and secondly to the Tokyo International Gift Show in 1990. The Association was delighted with the positive response to its jewellery items shown by the Japanese.

5. In 1991, another Association was organised by MJIA for the purpose of extending financial assistance to its members. It was registered under the Mutual Guarantee Association, created by the Cottage Enterprises Finance Programme of the Development Bank of the Philippines, the goal of which is to give the borrower access to credit financing with minimum collateral.

6. A co-operative was also organised by Ms. Ramos, called the Meycauayan Jewellery Industry Multi-purpose Co-operative, Inc., whose objectives and purposes are: *a)* to create funds to enable its members to secure loans for production purposes; and *b)* to provide goods, services and other requirements to its members, whenever necessary.

7. The major activity of the Association for 1996 was to develop and construct the Philippine Jewellery Centre, a multi-sectoral project of MJIA, in close collaboration with the Technical Education and Skills Development Authority, the Department of Trade and Industry, the local government of Meycauayan, and the Fourth Congressional District of Bulacan. This project will cost about P 42 million and is a shining example of a total collaborative effort between the government, local, provincial and central, and all the players in the industry.

Indeed, the women-led SMEs in the jewellery sector are doing their utmost to attain their common vision for the future – a thriving, vibrant jewellery industry, properly tooled up with increased capability to deliver high-quality products, not only for the domestic market but also for niche markets abroad.

Women in SMEs: roll of honour

Over the last 50 years, a few pioneering women have started their businesses from humble beginnings and successfully brought their companies to fame and fortune through hard work, sheer determination and excellence in production, marketing and exports. Some have even gone public – which is quite rare in the Philippines. What is most notable about these companies is that the second or third generation are already on-board, leading, or managing the businesses and they are all women! The following are some of the extraordinary women on our roll of honour:

Engracia Reyes, restaurant operations, processed foods – Carmencita Reyes (daughter) and Clara Reyes Lapuz (grand-daughter)

Cornelia Sy, hardware store operation and related businesses – Regina Sy Coseteng (daughter)

Trinidad Enriquez, restaurant, catering, hotels and resorts – Erlinda Enriquez Panlilio (daughter) and Katrina/Nicole Panlilio (grand-daughters)

Gliceria Rustia Tantoco, department stores, supermarkets, specialty stores – Nedy Tantoco/ Marilou Pineda/ Carmencita Lopez/Maritess Tantoco (daughters)

Socorro Ramos, bookstores, publishing and related businesses – Fred Ramos (son) Virgina Ramos (daughter-in-law)

Salud Tesoro, handicrafts, souvenirs and gift shops – Sally T. Guerrero (daughter)

Lucy Que, garments manufacturer – Cristy Q. Lee (daughter)

Elena S. Lim, prawn and consumer electronics exporter, industrial park developer – Susan L. Tan (daughter)

Carmen Yulo, manufacturer of novelty items and home decorations, real estate developer – Elena Y. Lorenzo (daughter)

Milagros S. Dayrit, jewellery manufacture and stores – Jacqueline Boncan and Michelle Soliven (daughters)

In Philippine society today, the above names are bywords and symbols in the industries and businesses they have started. For instance, the Aristocrat restaurant chain of the Reyeses, the D & E establishments (Sulo Hotel, Philippine Village Hotel and Puerto Azul) of the Enriquezes and Panlilios, Miladay jewelry of the Dayrits, Camelray of the Yulos, National Book Stores of the Ramoses, Rustan's Department Stores of the Tantocos, and Tesoro's Handicrafts are all well-known and popular. The Philippines is indeed blessed to have such enterprising women who started it all when it was not yet fashionable and who have risen to become true "capitanas" (captains) of their businesses, quite ahead of their time. We are very proud of all of them!

Concluding remarks

SMEs – and particularly women in SMEs – will have to face major upheavals in the near future: these include globalisation, rapid technological change and the liberalisation of trade and investment policies. While opportunities appear to be opening up in the region, keen competition will prevail, especially in the near term. This will mean that the shape and pace of business will also change – Philippine SMEs will have to become not only faster, but also smarter.

However, there is some good news ahead for women in SMEs. In the Philippines all the economic signs are looking up. In particular, exports are on the rise, tourism grew by 100 per cent in just three years, and markets beyond its borders, especially in East Asia, are all on the upswing. The Philippine government has also been providing a great deal of encouragement, beginning with its declaring last year as "SME Development Year". All these factors will favour the development of Philippine SMEs, in general, and Filipino women in SMEs, in particular.

Many of the problems that have prevented SMEs from being on the growth track in the past are now being resolved. The government, private sector agencies and institutions involved in SME initiatives have identified eight factors critical to SME success. These are: entrepreneurial education, financing, marketing, human resource development, science and technology, information dissemination, countryside infrastructure and institutional strengthening. As a result, these organisations have now redirected their efforts towards creating a more dynamic and supportive environment for SMEs.

SMEs are being encouraged to promote themselves and learn the basics of competition in the home market. They are being provided with opportunities for subcontracting in order to familiarise themselves with the merits of being partners with large corporations before making the big leap to exporting. Special incentives and assistance schemes for SMEs are being also made available, together with a number of special financing programmes through government development banks or similar agencies.

All of the above factors will certainly improve the climate for all the women in SMEs, although, clearly, much remains to be done. Nonetheless, it is heartening to note that the Philippines owes its current state of economic buoyancy to its women in SMEs, who have done more than their share in innovating and creating jobs in the countryside.

BIBLIOGRAPHY

APEC (1996), *APEC Business Advisory Council Report to APEC Economic Leaders.*

BONNIN, Marissa (1997), "Association of Negros Producers" report.

BUREAU OF SMALL AND MEDIUM BUSINESS DEVELOPMENT (1966), *Small Business in the Philippines*, report prepared by the Trade and Industry Information Centre.

CENTRE FOR INTERNATIONAL TRADE EXPOSITIONS AND MISSIONS (1997), Golden Shell Awards List of Recipients prepared by Ms. Elvie Borje.

LICUANAN, Victoria S., *Breaking Barriers: Businesswomen of South-East Asia*, Asian Institute of Management.

RAMOS,Cecilia R. (1997), "Meycauayan Jewellery Industry Association" report.

UNIDO, *Women in Manufacturing: Patterns, Determinants and Future Trends*,.

22nd Philippine Conference Handbook in Observance of the 1996 SME Development and Information Technology Year.

Chapter 14

THE DIALOGUE BETWEEN THE OECD COUNTRIES AND THE TRANSITION AND DEVELOPING ECONOMIES: AN INDIAN EXPERIENCE

by
Uma Reddy
Chief Executive Officer, MD Hitech Magnetics
India

India has realised that its commitment to the objective of economic growth with social justice can be fulfilled only if women are brought into the mainstream of economic activity. One of the most significant developments of recent years has been the promotion of entrepreneurship among women. Although the security, continuity and safety provided by an organisational shelter are important for most women, others have begun to view entrepreneurship as an economic and gainful option. Of a total 2 million small-scale industrial (SSI) units, less than 10 per cent are owned by women – the move by women towards entrepreneurship is still slow. The new industrial policy for small-scale industries announced by the Government of India in 1991 emphasized the promotion of entrepreneurship among women. The number of women entrepreneurs has not reached the critical mass necessary to make any impact on the system in either absolute or relative terms.

[In the context of globalisation], in order to ensure their survival, let alone sustained growth, Indian women-owned enterprises need to become more competitive and able to deliver quality goods and services at market rates without the help of incentives.

Global and regional networks with members committed to helping one another can enable women-owned enterprises in accessing markets, technology, and even capital, certainly in the developing economies, but possibly also in the transition and OECD economies.

In the two decades which have passed since the International Year of Women, women's movements across the world have focused on the economic development of women, with entrepreneurship development gaining attention in the early 1980s, even though India had served as an experimental ground for David McLelland during his famous Kakinada experiment in motivational training. Since then, a considerable number of women have ventured into entrepreneurship – the majority in micro and small enterprises.

Women face many constraints in setting up and running an enterprise. While these constraints vary from one country to another, there are many similarities. Efforts have been made in different countries to overcome these constraints, but these have been limited to some degree by lack of information and exchange of experiences between different countries. Sharing cross-country experiences on similarities and differences, on constraints and opportunities, can be fruitful in developing useful recommendations for intervention at the policy level.

Dialogue between countries is a key strategy. This paper looks at the situation from the Indian context, and at how such a dialogue can be sustained and utilised.

Introduction

In recent years, the world has made tremendous economic and technological progress. However, despite this economic progress, more than one billion people in the developing countries are struggling to survive, *i.e.* because of rapid population growth the poor are becoming more and more numerous in absolute numbers,. The number of people living below the poverty line in India exceeds the entire population of Africa. Hence, although in recent years the growth rate of the Indian economy has been around 7 per cent per annum, this has not resulted in any significant improvement in either GDP or quality of life for the majority of the poor. While the same processes cause poverty among men and women, women tend to be more vulnerable and find it more difficult to rise above the subsistence level.

Efforts towards economic development need to address the gender dimension of poverty as a priority issue. Poverty alleviation policies today look at special policies and packages for women in addition to their focus on economic growth. The emphasis on the promotion of women-owned small and micro enterprises is one such measure.

With the recent trends towards globalisation, there is greater emphasis on competitiveness, placing an additional burden on the micro enterprises in the country. It is therefore becoming imperative that adequate support be made available to these micro and small enterprises in order to improve their performances and their capability to withstand competition.

Gender and industry – a global overview

Women throughout the world make a vital contribution to industrial output. While the proportion varies from country to country, this contribution is not fully acknowledged and the status of women in employment is still not equal to that of men. The glass ceiling phenomenon has been sufficiently documented as to not require further elaboration.

It is in this context that entrepreneurship development among women assumes far greater significance. Women as business owners tend to employ other women and also tend to promote them to higher levels if they show the ability. Entrepreneurship and employment among women are governed by many socio-political factors. While in most South Asian countries the participation of women is lower in formal employment, interestingly this is not so in the South-East Asian countries. However, in the informal sector the presence of women as owners of micro enterprises is substantial in all these countries. It is these enterprises that face major constraints with regards credit, legitimacy and security.

Entrepreneurship among women in India

India has realised that its commitment to the objective of economic growth with social justice can be fulfilled only if women are brought into the mainstream of economic activity. One of the most significant developments of recent years has been the promotion of entrepreneurship among women. Although the security, continuity and safety provided by an organisational shelter are important for most women, others have begun to view entrepreneurship as an economic and gainful option. Of a total 2 million small-scale industrial (SSI) units, less than 10 per cent are owned by women – the move by women towards entrepreneurship is still slow. The new industrial policy for small-scale industries announced by the Government of India in 1991 emphasized the promotion of entrepreneurship among women.

The number of women entrepreneurs has not reached the critical mass necessary to make any impact on the system in either absolute or relative terms. Self-employment activities by women in poor and urban households remain outside the purview of mainstream productive activities, relegating them to the informal sector.

A typical woman entrepreneur's enterprise is very small, specialising in traditional manufacturing activities such as food and garments, with a low turnover, a minimum number of employees and no professional assistance. The subtle, but effective, manifestation of the gender phenomenon is reflected in the size of the business, its product line, growth, composition and management. This can be interpreted in terms of identity and expectation, vision and motivation, etc. The female psyche, combined with limiting social, economic, educational, financial and legal factors, together with the cultural milieu, contributes to the gender bias found in industry today.

In the informal sector most micro enterprises are self-employment ventures, bent on day-to-day survival, and not intended as a means of growth. Women manage petty trade or small food processing activities, often at bare subsistence levels, with any surplus being used for daily consumption. Any increase in incomes for these women therefore needs to focus on strategies for growth and adequate support and facilitation measures.

Issues facing women-owned enterprises – start-up, survival and growth

Alhough the three major stages in the entrepreneurial process – *creating, nurturing and nourishing* – are the same for both men and women, in practice the problems faced by women are of a different magnitude. This is due to social, cultural and economic reasons; the gender discrimination that prevails at all levels of society inhibits women's growth and development. Gender-based inequalities are such that women-owned enterprises face additional constraints, as can be seen below.

Credit

Accessing credit for starting an enterprise is one of the major constraints faced by women entrepreneurs, especially women in low-income brackets. General lack of experience and exposure restricts women from venturing out and dealing with banking institutions. Those who do so find that the transaction costs for accessing credit are high and cannot be met by the cash at their disposition.

In setting up her business to manufacture quilts for an export order, Rita George, after providing as collateral land worth Rs 1.35 million together with her husband's personal guarantee, received a term loan of Rs 0.2 million for fixed assets from Karnataka State Financial Institution. For her working capital requirement of Rs 0.4 million she approached a nationalised bank, but was turned away because she could provide no collateral. Even with bank-to-bank letters of credit, and three consignments worth Rs 0.3 million ready for delivery, she needed credit to meet delivery expenses, and was forced to apply for private financing, which cut into her profits.

Commercial banks are sceptical about women's ability, and insist on collateral security. Most women are resource-poor, particularly in terms of land and capital, and hence are unable to offer any collateral of their own. Therefore, they are dependent on family members (male) for surety or collateral – which restricts the amounts of money they can borrow. This results in lower investments. Alternatively, they can obtain working capital, but at much higher rates of interest. All these factors affect the enterprise and its survival. This is true not only for micro entrepreneurs, but also for larger, formal small-scale enterprises. Even here, women are only able to source credit on the basis of family assets, contacts, etc. Private financing is only possible at extremely high rates of interest.

In addition, the poor employment opportunities and low level of employment result in women's earning capacity being very low, they have limited savings and a low equity base to access credit. It is in this context that micro-credit, replacing the conventional collateral requirements by group liability, has proved to be a significant tool. Savings and credit associations formed by women's self-help groups have supported many an economic project, with loans being taken out either to improve the running of existing businesses, or to start new ventures.

One example is the module promoted by PRADAN, an NGO in India. The organisation promotes savings among the members of the self-help group right from start-up. Each group, which has 15 to 20 members, has its own code of conduct and regulations. Members of the group decide the amount to be saved, periodicity of saving, purpose of or priority of loan disbursement to members, rate of interest, etc. Each group has its officers who manage and take care of the accounts, hold and record minutes of meetings, maintain records, etc. PRADAN initially provided support to the groups through training in record keeping, account maintenance, leadership, etc., so that the SHGs are able to work independently and are self-sustainable. As soon as the SHGs have sufficient funds they start lending, almost immediately. Since funds are limited, PRADAN supports the SHGs by providing a matching grant to build the motivational level of the SHG and to expand the group resource base. Many SHGs have now federated at the cluster level, called "NIDHI", whose main function is to allocate funds sourced from the groups' share, interest earnings and bank loans. Several NIDHIs then combine to form a "Federation", which is a registered society and essentially facilitates linkages with institutions for leveraging credit. This three-layer method, created by PRADAN, provides a second higher layer to the primary savings and credit mechanism in recognition of the fact that SHGs are unable to mobilise sufficient capital to finance economic activities continuously. PRADAN has facilitated provision of up to 3 057 loans, amounting to some Rs 7 120 017 (US$2 million).

Similarly, other initiatives exist for providing alternative sources of credit:

From the NGO sector, SEWA – Self Employed Women's Association – set up a bank of its own. The majority of SEWA loans are unsecured, supported only by a reference person and investigation into the background of the applicant. The SEWA Bank encourages savings, as a women's need for savings is greater than her need for credit. By 1989 SEWA had deposits totalling Rs 15.6 million (US$4.5 million).

Other initiatives have come from organisations such as the Bridge Foundation which supports individual women entrepreneurs for loans up to Rs 8 000. Government policies have enabled banks to provide onlending to self-help groups. Rashtriya Mahila Kosh also provides funds to NGOs for onlending at low rates of interest.

As a result of these actions, there has been a substantial increase in the number of self-help groups engaged in savings and credit activities. This has in turn led to a large number of micro enterprises being started by women. Bankers admit that repayment by women is higher than that by men – at around 98 per cent.

It is interesting to note that it is self-reliance that motivates women to venture out on their own, but they have to rely on the securities of others to raise their basic capital requirements!

Social constraints

Many social factors make it more difficult for women to venture into gainful employment or entrepreneurship. First and foremost is the attitude that a "women's place is in the home", and that her first priority should be marriage, home and family. Girls are brought up with this attitude, and therefore few view employment as a valid option. Even for the few girls who are interested, the fact that marriage results in geographic displacement makes serious entrepreneurship a difficult option, usually considered only by older married women once the children have grown up.

Women from poor families invariably try to find some form of work. Household enterprises and micro enterprises are an attractive option, if the opportunity is available, as the women can usually take their children along to their place of business, and manage to attend to the needs of the children even in the midst of routine business activity.

However, women tend to start very small enterprises, due to lack of skills, lack of capital, lack of social approval and, often, lack of sufficient confidence in themselves.

Women entrepreneurs drastically limit the size of their enterprises as an insurance against failure. The fear of failure among women is very high. At the same time, women's own psyche and fear of success inhibits growth – success means a larger business and devoting more time to the business, therefore less attention to the family, which is her priority. Second, if she succeeds beyond a certain point, she might become a threat to her husband's ego. Rather than endanger family relations, women choose to level off at a certain comfortable zone, once having established their identity and proven their ability in their social and family circles.

In many traditional families, a woman taking out a loan represents almost a cultural revolution. When Ms. Vijaya Venkatesh announced triumphantly that she had obtained a term loan to set herself up in the manufacture of note books, the news was met with disapproval by her in-laws, and she was asked to reimburse her debt immediately. Vijaya sold off her personal gold and repaid her loan in order not to start her business on a note of confrontation. However, she continued to run her business. She was able to take care of her baby *and* manage the business. Her sister, also an entrepreneur, was her source of inspiration. Recently, she has taken out a loan to expand her business without having to face any disapproval, as she has been able to win the confidence of her family and destroy a few myths at the same time!

Attitude

The bias against women in business is strong because of cultural and traditional values. The behavioural pattern of women, characterised by modesty and lack of articulation, is often misinterpreted as incompetence and lack of professionalism. As it is, the transition from non-professional to professional status represents a formidable effort for a woman. Also, because they have been trained to listen, obey and leave the decisions to the men in the family, women find it difficult to interact and discuss issues as equals. The condescending views of others do little to support them.

Sandhya, a mechanical engineer, joined her father's business of manufacturing machines for the coir industry. The factory staff found the change in her position from owner's daughter to female professional on the shop floor difficult to accept. In addition, gaining acceptance from her clients, who continued to request that she send her father to discuss business, was a difficult task. A professional attitude, combined with perseverance and technical support to improve productivity, helped Sandhya gain the confidence of the staff and they have subtly changed their perception of her to that of a female professional.

What emerges from the above is that women are viewed not as business persons, but as women. *Interfacing and interacting* with women as professionals is still an awkward experience for many men due to the deep-rooted traditional perception of the male-female relationship.

Welfare vs. development

The approach taken by government agencies to entrepreneurial development for women is one of welfare, *i.e.* bestowal. This "by women, for women" approach keeps women from integrating mainstream activities. Some concepts – such as reserving certain items for women; exclusive industrial estates for women; exclusive training programmes; and employing 50 per cent women –tend to stunt rather than open up perspectives for women entrepreneurs.

There is an all-pervasive "gentle gender" phenomenon in the sphere of women in industry, a cumulative effect of psychological, social, economic and educational factors that act as impediments to women entering the mainstream.

Lack of resources/experience/skills/exposure

Women usually have fewer resources than men and are hindered by lower levels of education and literacy so that they are unable to qualify for jobs. In India, the literacy rate is 63.9 per cent for men and only 37 per cent for women. In addition, their low levels of remuneration mean that families prefer to educate boys.

Land rights are closely tied to inheritance and marriage laws. In a developing country such as India, where the main sources of income are derived from natural resources like land, the problem of lack of access to land is particularly critical for women, and further curtails their access to credit to venture into any income-generating activity where land is used as collateral to access credit.

Lack of access to education and training opportunities means that women start their businesses without the adequate skills. In traditional systems mothers transmit their know-how and skills to young women, but these are essentially an extension of women's domestic responsibilities such as stitching, food processing, etc. Most of the technical training offered to girls at the post-school level, in the women's polytechnic for instance, is limited to traditional careers such as secretarial practice, dress designing, etc. The exclusivity of training acts as limiting factor!

Products made at the micro-enterprise level tend to be of lower quality. Even if the quality is adequate, it is difficult for women to discuss technical details with customers. Hence, they tend to obtain lower prices for their products. It also appears that women tend to locate their businesses more from the point of view of distance from home than from that of "access to a local market".

A typical example is the State of Karnataka, in India, which for the last 200 years has been a traditional silk-producing state. Only the men are involved in the marketing of cocoons, yarn or woven cloth. While women are very much part of the production processes, marketing is perceived to be a male domain. The male-only environment in the market further inhibits women from joining them.

Lack of networking

Women have fewer contacts, less idea of how to deal with the bureaucracy and less bargaining power – all factors which further limit their growth. Since most women entrepreneurs operate on a small scale, and are generally not members of professional organisations nor part of a network, they find it difficult to access information. Most of the existing networks are male-dominated and women feel uncomfortable in them. These networks, in turn, do not welcome women and prefer to remain exclusive. Even when a woman does venture into one of these networks, her task is made more difficult because most networking activities take place after regular working hours. There are very few women-only or women-majority networks into which a woman could enter, gain confidence and move on. This lack of networks deprives women of awareness and exposure to good role models.

> A significant exception to this situtation is "AWAKE", where the 300-odd voluntary entrepreneur members provide a platform for women to have a peer-group network. Women from diverse backgrounds provide role models for potential women entrepreneurs. They are able to support each other through the sharing of experiences, knowledge, information, suppliers, etc. Experienced women hand-hold new members and improve their confidence. Members say that AWAKE gives them an identity as an entrepreneur.

Fostering growth

It can be seen that women face many barriers in starting their enterprises, surviving and growing. While credit – or rather lack of credit – is a very important factor, it is not the only one. It is also clear that even where credit has been made available, in the absence of other facilitating factors the enterprises created tend to remain at subsistence level, and do not grow.

> The question raised here is: What innovative mechanisms can be used to sensitise women and to foster innovation among women entrepreneurs so as to enable them to grow? An upward dynamism is required, as follows.

Examples of efforts to encourage women entrepreneurship

Organisations

There is need for a catalyst to motivate women into taking up entrepreneurial activities by making use of the available opportunities while simultaneously sensitising implementing agencies to the requirements of women entrepreneurs.

For women, an organisation is a platform for building self confidence, a means for breaking out of the confinement of the home. Several associations and organisations help entrepreneurship development in the small-scale sector. While these organisations often have women's wings or cells, those exclusively engaged in promoting entrepreneurship are very few in number. In fact most women's groups all over the world are engaged in credit provision.

AWAKE – Association of Women Entrepreneurs of Karnataka – is one of the pioneers in the field of providing non-credit support. Thanks to the vision of its founding president, Madhura M Chatrapathy, while a number of organisations have taken up the cause of women, AWAKE provides peer-group support and hand-holds new entrants in various aspects of entrepreneurship. Apart from motivating potential women entrepreneurs, it conducts growth-oriented programmes aimed at sustaining the businesses of existing entrepreneurs. The business ventures of the 300-odd voluntary members of AWAKE range from manufacturing of electronic components, injection moulded components, machinery for the coir and silk industry; through trading in fruit juices, stationery, etc.; to using AWAKE's business incubator for food processing. It should be mentioned here that AWAKE is an affiliate member of Women's World Banking (WWB), but is not engaged in credit provision. The unique feature of AWAKE is successful entrepreneurs helping upcoming entrepreneurs.

After attending AWAKE's EDP programme, Ms. Malathi Hebbar and Ms Vijaya set up a self-service, fast-food place - "Vybhav Darshini" . As part of the programme, the then Vice-President Padma Seshadri, who ran a hotel, shared her experience with the neo-entrepreneurs and became their role model. They spent a few days in Padma's hotel learning the tricks of the trade. Thanks to the efforts of AWAKE in providing lectures by faculty from banks and government departments, the girls did their networking with the bank to obtain a loan. Although the training triggered their venture, it was the continuous hand-holding which supported them and kept their spirits up.

AWAKE looks at entrepreneurship from a holistic point of view rather than as a simple component of training. From a study of the business counselling sessions of AWAKE, it was found that a significant number of women wanted to start enterprises in food processing. To facilitate this, AWAKE set up its business incubator, the first of its kind in Asia. Here women can make use of the machinery, FPO licence, consult withf the food technologist, etc., to manufacture and standardize their products. Incubator users, Ms Indira Channegowda and Ms Jyothi Murthy have incubated and set up their own enterprises. AWAKE also intervenes to support entrepreneurs in accessing formal credit. In the process, follow-up, refresher courses and growth programmes are just as important as start-up.

Business guidance, moral support, confidence building, problem solving and help in negotiations are all available through peer-group support. These links with *best practitioners* who act as mentors and counsellors are part of the support system. Such interventions reduce the chance of micro-enterprise failure by providing opportunities for close monitoring and immediate corrective action in case of problems. *Role models* can provide inspiration, motivation and an opportunity for learning.

An example of a government initiative to bring a co-ordinated effort to the development of trade-related entrepreneurs among women is TREAD, which largely focuses on micro enterprises. While a large number of women have been engaged in trade-related activities for a long time, they do not have a very visible role. Retail trading, generally termed petty trade, has always been a special forte of women. "Hat", push-cart, house-to-house and street-side vending by women is a common sight in India. In addition, a large number of women run boutique-style, up-market trading outlets.

We also have a number of womenactive in export marketing as merchant exporters. To increase the number of women in trading and trade-related activities, especially export trade, UNDP, GOI and ITC Geneva have initiated a scheme to involve 100 000 women in Trade and Trade-related Entrepreneurship Development – TREAD – over the next five years. As a part of the programme, export villages/clusters are being created. This provides a very large scope for co-operation.

Some organisations such as small-industry associations, trade associations, chambers of commerce and sector-specific associations, act as pressure groups and lobby with the government for special concessions, subsidies, etc. Some of these associations have women's wings to cater to the needs of women members. Even some banking and financial institutions now have women's cells to cater to women's needs. In Karnataka, both the State Financial Corporations and the Government District Industries Centre have women's cells in all their district-level branches.

Role of grassroots NGOs

It is estimated that there are over 40 000 NGOs in India, most of which target women as an integrated part of their programmes to ensure sustainability of their projects. NGOs have extension officers located near villages, who are in direct and regular contact with the women and who therefore understand the relevant issues in the local context. They have skills in group formation and capacity building. Given their contacts, networks and linkages, they can help the women in the villages overcome illiteracy, can motivate them and can liaise with the entrepreneurship development organisations to determine the appropriate income-generating activities.

> Astha Sansthan, an NGO in Udaipur district, Rajasthan, in 1986, provided support in forming a co-operative to the tribes who collect tendu leaf (beedi leaf). The work involved leaf collection and packing the leaves into bundles. The women felt that they were not sufficiently paid for their work. Astha Sansthan supported the tribal groups in estimating their costs. The middleman, *i.e.* the contractor, was thrown out and the co-operative learned the procedures necessary to become a "contractor". It was a mixed gender group, with women making up almost half of the members. Women did most of the collection work and in addition took an active part in quality checks, motivating other tribal groups, taking an active part in meetings and financial discussions, etc. They earned a net profit of Rs 700 000 in 1991.

The capacity of organisations to negotiate and to direct development processes on their own in a sustainable way is rooted in the combined effort of a strong organisational base and negotiating capacity and the focus of attention on women's issues. Such organisations could support women's groups, thus giving them collective bargaining power. They could also support the groups in market identification, market penetration, as well as organising buyer-seller meetings, etc.

Role models

Women wishing to set up their own business are inhibited by a lack of role models. One-off stories of women achievers in high-tech industries are not sufficient to enthuse the general run of aspiring women entrepreneurs.

> AWAKE constantly promotes the family as an institution in order to encourage family members to realise the need to support women's efforts. Meera and Kalyani, while sharing their business experiences with a group, talked about their family and the self-reliance of their children. "Meera, who runs a fresh fruit juice vending business, has successfully raised her son to be an engineer. Kalyani, who runs a book-manufacturing company, has her son in the army." One of the participants, Ms Shahnaz, wanted to start up a florist shop, but her husband was against the idea because he felt it would take time away from the children and the support they required for their studies. After listening to Meera and Kalyani, he was convinced and was willing to support his wife's endeavours. Meera and Kalyani became Ms Shahnaz's role models. AWAKE has several down-to-earth entrepreneurs whom they use as role models to encourage entrepreneurship among women.

The two most important things for women entrepreneurs are a "mentor" and a "network" – a forum where women can communicate as equals, share ideas, make contacts and do business with each other. This network should extend beyond state and national borders.

Even government staff can motivate and empower women:

> A lady Deputy Commissioner in Tamil Nadu empowered a group of women by giving them a licence for quarrying. The women were allowed to quarry directly and were thus out of danger of becoming bonded labour to contractors.

Gender specific counselling/training – the AWAKE model

Providing business counselling on an ongoing basis reduces the problems of start-up, as well as established, women entrepreneurs. AWAKE has been working for over 12 years in the field of support for women entrepreneurs. It soon became apparent that the standard Entrepreneurship Development Training Programmes were not appropriate and that new programmes needed to be developed. Since that time AWAKE has been at the forefront of designing and conducting appropriate training programmes for EDP.

The first step was to provide business counselling on a regular basis. AWAKE provides free business counselling for potential as well as established members by voluntary member entrepreneur counsellors every Thursday afternoon. The counselling sessions, in addition to providing business guidance, also provide an opportunity for potential women entrepreneurs to meet their models and exchange ideas on business opportunities. These sessions have also helped AWAKE to compile data from which a unique training model for women entrepreneurs has evolved.

Data gathered from the counselling sessions gave an insight into the specific problems faced by women and the areas in which further support was needed. Two different levels of training were developed for different levels of entrepreneurship. One feature common to all the programmes was the introduction of an empowerment module, and the use of a core faculty or trainer who would be available for the duration the programme, irrespective of who the guest faculty might be.

Another important factor was the extensive use of members of AWAKE as role models with whom the trainees could strongly identify. In addition, efforts were made to bring in as guest faculty, officials from infrastructure, credit and other development agencies, as a start to future contacts and networking.

Initially AWAKE focused in Bangalore as its reach was limited. With the growing awareness and recognition of AWAKE's efforts, the demand for conducting similar programmes in other towns and rural areas increased, including requests from the state government. However, it was realised at AWAKE that one-time training alone would not lead to success and constant hand-holding was essential. AWAKE could do this in Bangalore city, where most of its members were located, but obviously would not be able to provide such support continuously elsewhere. From this arose the idea of using local organisations and NGOs to work in partnership with AWAKE so that the NGOs would be involved in the training, counselling and subsequent hand-holding process. This is currently being put into practice in the RIP for Bijapur district, supporting the Government of Karnataka and SIDBI.

As the concept of entrepreneurship was unfamiliar to most NGOs, AWAKE had to develop the initial contacts, persuade the NGOs to work with it, and then develop and level of support that AWAKE had been providing.

Training support – co-operation between business and NGOs

It is observed that while governmental organisations have more difficulty in reaching women entrepreneurs, NGOs have difficulty in conducting effective programmes due to a lack of business experience. In such cases NGOs could link up with a corporate body for technical and marketing support.

One such example is the project implemented by the Kasturba Gandhi National Memorial Trust – an NGO, which, together with the support of the corporate company M/S Ponds (India) Ltd, is training rural women in the manufacture of leather footwear uppers with the objective of supporting income-generating activity for rural women. M/S Ponds provides the women with jobwork in the manufacture of shoes. This equips the rural women with a skill which will support them financially.

Networking for growth

The sustainability of networks requires mutual long-term interest in the form of buyer-seller links, consortia formation, etc. However, before the network reaches a level of maturity where the different actors are capable of developing such links, much work needs to be done over a number of years, during which the network has to be kept alive by outside support. Typical activities in the formative period would be training, competence building in specific sectors of business, developing product portfolios, electronic marketing links, etc.

In the context of women, technology is perceived as anti-women since it is felt that machines displace women from the workforce and that women cannot handle many of the technologies – *i.e.* machines. Technology is related to women only in the context of reducing drudgery and not as a tool for economic development. On the other hand, simple techniques and tools, processes and equipments could help in business creation. However, little effort has been made in this field. Ms. Madhura M Chatrapathy of ACE perceived a great opportunity for support to women in this area and has made a substantive contribution in taking this idea forward, particularly in intra-regional Asian and Pacific region and Afro Asian co-operation through efforts of technology sharing programmes for women particularly for the benefit of rural and grassroots women's groups.

The first major effort was to identify technologies in eight Asian and Pacific countries and support creating facilitators and a network of support agencies in participating countries. The Facilitators Handbook, an output of the project, is in great demand. The project was that of UN-ESCAP APCTT and was supported by the Dutch Government.

Currently another five-country project is being implemented by APCTT with ACE – Asian Centre for Entrepreneurial Initiative – to bring together grassroots women's groups and engage in food processing to share do-how, know-how and technology transfers. There is need for further efforts in this area to provide technology as a tool for enterprise development. There is need for tech banks and tech networks for women.

While a large number of government and private sector agencies exist at the district, state and national levels, systems for access and linkages need to be established with these organisations if the women's groups are to take advantage of the technical and marketing support that they can provide.

Globalisation – challenges and opportunities

The onset of liberalisation in India has led to many changes in policy, most of which tend to place small and micro enterprises at a disadvantage. The risk that social sectors and needs will be marginalised in a market-driven and export-oriented world will touch women disproportionately. Women-owned enterprises fall into this category, and can be expected to be among the worst hit. In order to ensure survival, let alone sustained growth, these enterprises need to become far more competitive and able to deliver quality goods and services at market rates without the help of any kind of ongoing incentive.

But the act of becoming more efficient involves, among other things, accessing better technology, accessing global information and developing skills in continuously identifying new products for global markets, especially in view of shortening product life cycles. All this costs a considerable amount of money – money which the women do not have.

It is in this context that global and regional networks with members committed to helping one another, need to be developed. Such networks can help women-owned enterprises in accessing markets, technology, and even capital, certainly in the developing economies, but possibly also in the transition and OECD economies.

However in order to yield mutually beneficial results, members of the network will need to undergo considerable capacity building, in order to become capable of accessing, understanding, absorbing and utilising the information available. This requires that sufficient time be provided for members from different countries to develop an understanding of the differing social/economic/ political ethos, and hence an understanding of one another and for trust-building. In this phase the network needs to be initiated, sustained and co-ordinated by an external agency – this phase may extend to over three to five years.

One example of such an effort is the Enterprise Learning Network of WWB, which has proved beneficial to the women in the garment sector. It has, however, still to reach the self-sustaining level for a majority of the members. However, there have been a few cases of success for some members.

An effort was made to establish a network of women entrepreneurs in Asia, by ILO SAAT with funding from SwedeCorp. It consisted of a five-country project covering training of trainers, production of a training manual, exchange visits, workshops, exchange of information through a newsletter. Many of the task-oriented activities were successfully completed. However, the establishment of a sustainable network could not be achieved because further funds were not available and participating countries themselves needed further support.

Conclusions and recommendations

For this we need to look at agencies in different countries involved in entrepreneurship development among women, beyond the provision of credit alone. Such agencies, if not already in existence, may need to be created and nursed until such time as they are able to provide the necessary support independently. The network would probably need to exist at different levels – global, regional, national and local; and involve a combination of different agencies and individuals – *e.g.* organisations such as grassroots NGOs, AWAKE, credit agencies, federations and international organisations such as WWB and others, so that the focus is on a holistic approach to the development and continued support of entrepreneurship among women.

There must be both visibility in the existing mechanism and a desire to discover new mechanisms; this will hopefully lead to the creation of interlinkages which will foster continuity in dialogue. Some efforts are already underway and there is scope for additional work by the various agencies involved. In my personal view, there are a number of action points:

The role of the OECD

◊ *Establishing learning partnerships*:

- Setting up a co-ordinating body for sustaining networks – it is possible that ILO, SAAT, UN ESCAP, APCTT efforts should be further supported in order to develop more local organisations as a prelude to mutually beneficial exchange.

- Co-ordinating efforts of other development agencies in already identified thrust areas, *e.g.* in India, UNDP is working on a country co-operation framework focused on jute, leather, etc. Synergising efforts in conjunction will be more effective. As will co-ordination with organisations such CBI, TOOL, ITC, etc., for the provision of training in exports, technology transfer, trading, etc.

- Analyse the experiences gained from other regions and learn from them how increased globalisation has affected women. Develop a regional information network with national centres where women can exchange information on trade opportunities and share their experiences.

- Periodic interactions/experience exchange workshops between network members from different countries, focusing on women entrepreneurs. Conferences of women should be used as fora to project entrepreneurship, and in which discussion could lead to effectively bringing down barriers.

- Formal and informal training including gender training.

◊ *Technology sharing networks*: Setting up information exchange, possibly using Internet, with access routes to all members of network, including building the capacity of network member organisations to have computerised centres capable of accessing:

- Technology sharing and transfer with do-how and know-how including incremental technology.

- Information on scientific and objective analysis of structural, social and economic changes and the implications of major changes in the role of women in economic development.

- Access to market information and best practices. To identify alternative income opportunities and provide the assistance needed by producers in order to better respond to the demands of the international market.

- Exchange of information on better ergonomical furniture and tools for women to support and improve their productivity. Making available second-hand machinery and hand tools for micro entrepreneurs.

- Information on intellectual property rights and environmentally friendly technologies.

◊ *Resource networks*:

- Data availability and the serious limitations of their comparability represent an important constraint at present. Due to national differences in economic structures, the same statistical measures may cover phenomena of different economic and social significance. Further research is therefore needed on the statistical side of women's role in the economy.

- Gender-disaggregated data should be collected, analysed and made available at all levels and for all sectors in a continuous and systematic fashion.

- More research is needed to assess which institutional mechanisms are most effective in dealing with various elements of new economic policies that have a direct impact on women.

- Develop educational and informative material to advise women of new product and project ideas, new trade policies and economic reforms and how they relate to their needs and concerns. Establish regional and national research mechanisms to review and monitor existing and new trade policies and their effects on women.

- Support to training activities in developing countries at different levels:

  - network organisation – to build and sustain networks;
  - collective organisations – improved marketing, technical and financial strengths;
  - individual entrepreneurs – start-up and growth training, special export promotion programmes and information technology skills (information collection).

- Role models – success stories of women entrepreneurs, information, etc., can be shared to facilitate the promotion of entrepreneurship development of women through the exchange of experiences and projecting the image of women entrepreneurs in different countries.

- Introduce mechanisms for continuous provision of information through Internet. This also applies to teleconferencing.

◊ *Creating innovative financial instruments*:

- Establishment of budget lines for women – credit.

- Venture capital funding for women entrepreneurs and joint venture support.

- Credit guarantee scheme for women exporters.

Role of governments

Working at government-to-government levels for greater policy support to overcome the barriers faced specially by women in:

◊ Infrastructure provided to women to reduce drudgery and hence give them time to access education, skill acquisition and income generating activity.

◊ Study and analyse the socio-economic problems associated with the impact of trade expansion relating to women in their multiple roles as producers, consumers or social reproducers.

◊ In light of the increased unemployment which has followed the recent economic reforms and liberalisation, a "social safety net" is a prerequisite to protect deprived women. Those sectors which are most likely to suffer from international competition should be targeted for financial and technical assistance. On the other hand, many new income and employment opportunities are emerging with liberalisation and women should be provided with the necessary skills to take full advantage of these openings.

◊ Promote export sectors where women predominate. With delicensing, create an environment to promote exports and healthy competition.

◊ Come to grips with the macro sociocultural and institutional forces that may facilitate or hamper entrepreneurship development in the state, and implement appropriate policies with a regional focus.

◊ Mobilise and train NGOs in the region to take an active interest in global trade issues and to campaign for policy changes and reforms on behalf of women's needs and concerns in order to promote entrepreneurship and provide economic information with regard to new opportunities.

◊ Innovations required in the areas of promotional and information material that reach women, programme content, timing and location that are adapted to women's educational and skill levels, and their time and mobility constraints.

◊ Innovations to overcome the constraints to women's access to formal credit through simplified forms and procedures.

◊ Investment in marketing centres which support display of products of women enterprises.

◊ Create an environment for sustainable development with support systems.

Role of NGOs

NGOs can intervene to foster entrepreneurship development by:

◊ Many women, as home-based workers or engaged in micro-entrepreneurial activity, need to be organised in order to benetif from alternative production, marketing and access to raw materials. Assist groups to become better organised and responsive to the needs of their members.

◊ Build marketing linkages with international organisations in identifying markets, such as alternate trade organisations.

◊ Through technology intervention, use the network to locate and support women to upgrade their production processes into higher value-added products.

◊ Conduct gender sensitisation training programmes for all concerned, including both government and non government agencies.

◊ Lobby for simplified trading procedures and rules for small-scale producers.

◊ Act as matchmakers and facilitators and tie up with other organisations for providing skill or managerial training for women.

◊ A multidimensional view of training and development considering the content of the training; the different stages at which training has to be provided to an entrepreneur and the system level for training inputs.

◊ NGO-NGO co-operation in sharing of experiences, technology and expertise.

Role of women

Women themselves can speed up the process of entrepreneurship by:

◊ Mobilising themselves into groups at the grassroots level, forming networks to support themselves and to facilitate access to finance, markets, training, information, and enable them to negotiate on better terms.

◊ Women should endeavour to be represented at the community level, in policy-making bodies as well as in other trade-related bodies.

◊ Women should work to intensify and expand economic development in order to reap the benefits of increasing exchange and mutual co-operation among themselves and other countries. Economic co-operation should be intensified and expanded.

Acknowledgements

My sincere thanks to the OECD for giving me the opportunity to attend this conference and present this paper.

My special thanks to Ms. Madhura M Chatrapathy and Mr. Ram Bhat, of ACE, for their input to this paper.

In addition, I would like to extend my thanks to Mr. Narsimhan and Mr. Sinha for their guidance.

NOTE

1. Ms. Uma Reddy is an electrical engineer running a small-scale enterprise manufacturing electronic transformers. She began her entrepreneurial career while at college designing PCBs. She is actively associated in entrepreneurship development among women and has been serving AWAKE – Association of Women Entrepreneurs – in Karnataka in various capacities for the past seven years. She is an Executive Committee Member of Consortium of Electronic Industries of Karnataka. She has participated in the Regional meeting of Women's World Banking in Tokyo and the Export Promotion programme of CBI, Netherlands. She has been a member of evaluation missions by UNDP, FACET BV for NOVIB-funded projects and other agencies, and has organised many workshops, seminars, etc.

BIBLIOGRAPHY

AWAKE, *Policy Practice Gap with Reference to Collateral Security*, A study by AWAKE.

P&P GROUP, *Woman and her Enterprise: A Study in Karnataka.*

IIM, *Development of Micro Enterprises by Women*, Ahmedabad.

INTERNATIONAL LABOUR ORGANISATION (1992), *World Labour Report,* Geneva.

INTERNATIONAL LABOUR ORGANISATION, *Gender, Poverty and Employment: Turning Capabilities into Entitlements*, Geneva.

IYER, Lalitha, *Women Entrepreneurs: Challenges and Strategies.*

MADDISON, Angus (1989), *The World Economy in the 20th Century*, OECD Development Centre Studies, Paris.

MURPHY, Josette L., *Gender Issues in World Banking*, OED.

PAUL, Samuel, John C. ICKIS and Jacob LEVITSKY, *Educating Managers for Business and Government: A Review of International Experiences.*

OECD (1971), *The Challenge of Unemployment to Development and the Role of Training and Research Institutes in Development*, Paris.

UNIFEM and SIDA (1995), *Proceedings* of the Regional Seminar on "Global Trading Practices and Poverty Alleviation in South Asia: A Gender Perspective".

OECD Conference

WOMEN ENTREPRENEURS IN SMALL AND MEDIUM ENTERPRISES

"A MAJOR FORCE IN INNOVATION AND JOB CREATION"

Paris, 16-18 April 1997

ACKNOWLEDGEMENTS

The conference was jointly organised by the Industry Committee's Working Party on SMEs (Directorate for Science, Technology and Industry) and the Local Economic and Employment Development Programme (Territorial Development Service) in co-operation with the Employment, Labour and Social Affairs Committee's Working Party on the Role of Women in the Economy (Directorate for Education, Employment, Labour and Social Affairs), the Development Assistance Committee's Expert Group on Women in Development, and the Development Assistance Committee's Working Party on Financial Aspects of Developmental Assistance (Development Co-operation Directorate).

In the preparation of the conference, the OECD Secretariat was assisted by Ms Virginia Littlejohn, President, Global Enterprise Group (United States) as special advisor, and a team of international experts and academics: Professor Bertrand Duchéneaut, Director-General, Groupe Ecole Supérieure de Commerce (ESC) and Director, Centre EURO PME (France), Ms Patricia Mahot, Associate Partner, Studio Entreprendre (France), Ms. Julie Weeks, National Foundation for Women Business Owners (United States), Dr. Mary Barrett, Sunshine Coast University College (Australia), Ms Sylvia Runeborg, National Board for Industrial and Technical Development (NUTEK) (Sweden), and Ms Claude Cattan, Istituto per lo Sviluppo della Formazione Professionale dei Lavoratori (ISFOL) (Italy).

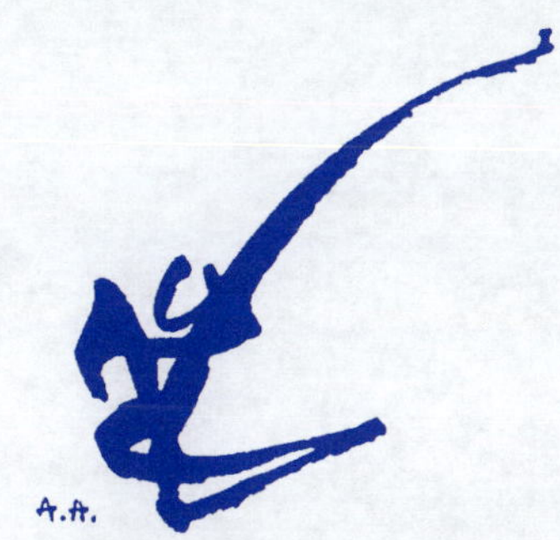

OECD CONFERENCE ON
"WOMEN ENTREPRENEURS IN SMEs
A Major Force in Innovation and Job Creation"
16-18 April 1997

CONFERENCE PROGRAMME

WEDNESDAY, 16 APRIL 1997

MORNING

9.30-10.30 — **Opening Session**

Opening Address:

Mr. Donald Johnston
Secretary-General, OECD

Speakers:

Mr. René Monory
President of the Senate, France

Ms. Anna Finocchiaro
Minister for Equal Opportunity, Italy

Mr. Heinrich Kolb
Secretary of State, Ministry of Economy, Germany

Ms. Ann Diamantopoulou
Secretary of State, Ministry of Development, Greece

Ms. Rita Dionne-Marsolais
Minister of Industry and Commerce, Quebec, Canada

Ms. Betsy Myers
Director of the White House Office of Women's Initiatives and Outreach, United States

Mr. Michel Boussard
Vice President, Global General Business Marketing & Business Development, IBM Euro Co-ordination

10.30-11.00 — **Cyber cafe pause**

11.00-13.00 — **Plenary Session: Women Entrepreneurs in SMEs: A Major Force in Innovation and Job Creation**

Chair:
Ms. Barbara Prammer
Federal Minister for the Condition of Women, Austria

Brief Introductory Remarks:
Mr. Arturo Gonzalez Romero
Chairman, OECD Working Party on SMEs, Spain
Mr. Carlo Borgomeo
Vice-Chairman, OECD LEED Directing Committee, Italian Agency for Youth Entrepreneurship, Italy
Ms. Joke van den Boer
President, Les Femmes Chefs d'Entreprises Mondiales (FCEM), Netherlands
Ms. Rajni Aggarwal
Secretary-General, International Federation of Women Entrepreneurs (IFWE), India

"Horizon Speaker":
Ms. Ronnie Feit
Rona Feit & Associates, United States

Presentation of Main Issues:
Mr. Thomas Andersson
Deputy Director, OECD Directorate for Science, Technology and Industry

Presentation by the Author of the Background Report:
Mr. Bertrand Duchéneaut
General Director, Groupe Ecole Supérieure de Commerce (ESC) and Director, Centre de Recherche Euro PME, France

Discussion by a Panel:
Ms. Laura Liswood
Director, Women's Leadership Project, United States
Ms. Leila Webster
Private Sector Development Specialist, The World Bank Group
Ms. Patricia de Müller
President, Asociación de Mujeres Mexicanas Jefes de Empresas, Mexico
Mr. Didier Adès
Journalist, France Inter, France
Ms. Nancy Barry
President, Women's World Banking, United States

Open Discussion

WEDNESDAY, 16 APRIL 1997			
AFTERNOON			
14.30-18.00 **16:15-16:35 Cyber cafe pause**	**Workshop 1: Issues in the Business Environment and Technology** ⇒ Representation for Women Entrepreneurs ⇒ The Changing Role of Associations ⇒ Technology and Innovation in Women-owned SMEs ⇒ Regulatory Issues for Women-owned SMEs *Chair:* **Ms. Carolyn Chin** Vice President, Strategic Projects, IBM Corp., United States *Speakers:* *US Case Study Keynote:* **Ms. Amy Millman** Executive Director, National Women's Business Council, United States **Ms. Sherrye Henry** Deputy Assistant Administrator, Office of Women's Business Ownership, Small Business Administration (SBA), United States **Ms. Suzanne Taylor** President, National Association of Women Business Owners, United States **Ms. Jocelyne Attal,** Vice President, NT Marketing, IBM Software Solutions Division, United States **Ms. Lorraine Martin** President, Lorraine Martin College, Australia **Ms. Kaori Sasaki** President, UNICUL International, Japan *"Horizon Speaker":* **Mr. Thierry Gaudin** Fondation 2100, France *Discussants:* **Mr. Charles Riley** Editor in Chief, WE Magazine, United States **Ms. Renée Richard** President, Belgian Association of Women Entrepreneurs, Belgium **Ms. Helen Appleton** Editor, Intermediate Technology Publications Ltd., United Kingdom **Ms. Louise Guay** President, Public Technologies Multimédia, Canada *Rapporteurs:* **Ms. Françoise Coré** Principal Administrator, OECD Directorate for Education, Employment, Labour and Social Affairs **Ms. Silvia Runeborg** Project Manager, National Board for Industrial and Technical Development (NUTEK), Sweden	**14.30-18.00** **16:30-16:50 Cyber cafe pause**	**Workshop 2: Globalisation and International Trade** ⇒ Issues and Opportunities ⇒ Financing Exports by Women-owned SMEs ⇒ International Investment ⇒ International Subcontracting *Chair:* **Ms. Virginia Littlejohn** President, Global Enterprise Group, United States *Speakers:* *Keynote speaker:* **Ms. Andrina Lever** President, Women Entrepreneurs of Canada, Canada **Mr. David Bowie** Deputy-Director, Office of Export Promotion, International Trade Administration, Department of Commerce, United States **Ms. Sarah Hage-Ali** Vice-President, Ghana Association of Women Entrepreneurs, Ghana **Ms. Angeline Low** Board Member, Australian Council of Businesswomen, Australia **Ms. Ingrid Andersson** President, PREFO AB, Sweden *Discussants:* **Mr. Jean-Marie Daillet** Former Ambassador, Honourary Member of Parliament, France **Ms. Jocelyne Gaudin** Head of Unit, European Commission, Directorate DG XX111-C1 **Ms. Eileen Cassidy** Assistant Administrator and Director of International Trade, Small Business Administration (SBA), United States **Ms. Guadalupe Gomez Fouz** Manager, Gomez Fouz S.L., Spain *Rapporteurs:* **Ms. Marie-Florence Estimé** Principal Administrator, Secretary of the Working Party on SMEs, OECD Directorate for Science, Technology and Industry **Ms. Marie-Sylvine de Lafforest** Consultant, OECD Directorate for Science, Technology and Industry

THURSDAY, 17 APRIL 1997

MORNING

9.30-13.00

Workshop 3: Management Issues

⇒ The Changing Nature of the Firm: The Role of Women-owned SMEs
⇒ Specific Features of Women's Management
⇒ Model Business Mentoring and Business Development Programmes for Women
⇒ Management Training of Women Entrepreneurs

Chair:
Ms. Gabrielle Rolland
Institut Européen du Leadership, France

Speakers:
Keynote speaker: **Ms. Candida Brush**
Assistant Professor, Boston University, United States

Ms. Maura Franchi
Head of Labour Policy Service, Bologna, Italy
Ms Francesca Santoro
Representative of an Italian Trade Union, Italy
Ms. Gina Zabludovsky
Professor, University of Mexico, Mexico
Mr. Salo Grabinsky
Consultant, Grabinsky Alvarez S.C., Mexico
Mr. Sachio Semmoto
Professor, Keio University, Japan
Ms. Naoko Banno
President, Career Strategy Inc., Japan

11:15-11:35 Cyber cafe pause

Discussants:
Ms. Leyla Khaïat
President, CNFCE, Tunisia
Ms. Arline Woutersz
President, British Association of Women Entrepreneurs, United Kingdom
Ms. Jane Applegate
The Applegate Group, United States

Rapporteurs:
Ms. Geneviève Lecamp
Principal Administrator, OECD Territorial Development Service/LEED

Ms. Julie Weeks
Director of Research, National Foundation for Women Business Owners, United States

9.30-13.00

Workshop 4: Dialogue among OECD Countries, Transition and Developing Economies: Future Challenges for Policy Makers

⇒ Exchange of Experiences and Review of Specific Situations
⇒ Co-operation among Countries

Chair:
Mr. Joop Vianen
Member, Board of Directors, KIT Royal Tropical Institute, Netherlands

Speakers:
Keynote speaker: **Ms. Swannee Hunt**, Ambassador of the United States to Austria

Ms. Susan Davis
President, Susan Davis International, United States
Ms. Uma Reddy
Secretary-General, AWAKE, India
Ms. Zenaida Gordon
President, Gordon Enterprises, Philippines
Ms. Florence Lardenchet
Association for Economic Initiative, France
Ms. Raluca Bejan
President, Association of Women Managers, Romania
Ms. Sonia de Avelar
CENPROFEM, Brazil

11:30-11:50 Cyber cafe pause

"Horizon speaker": **Ms. Nancy Barry**
President, Women's World Banking, United States

Discussants:
Mr. Olavi Änkö
Director General, Ministry of Trade and Industry, Finland
Ms. Lois Stevenson
President of Canadian Council for Small Business and Entrepreneurship, Canada
Ms. Etta Carignani
Vice President, FCEM, and National President, AIDDA, Italy
Ms. Anikó Soltész
Director, Seed Foundation, Hungary

Rapporteurs:
Ms. Francesca Cook
Administrator, OECD Development Co-operation Directorate

Ms. Mary Barrett
Senior Lecturer in Management, Faculty of Business, Sunshine Coast University College, Australia

THURSDAY, 17 APRIL 1997

AFTERNOON

14.30-18.00

Workshop 5: Financing Issues

⇒ Start-up Financing and Microlending
⇒ Financing for Business Growth: Mid-level (Mezzanine) Financing Issues

Chair:
Mr. François Beaudoin
President and Chief Executive Officer, Business Development Bank of Canada, Canada

Speakers:
Keynote speaker: **Ms. Patricia Mahot**
Associate, Studio Entreprendre, France

Mr. Henry Savajol
Director of Research, Crédit d'équipement aux PME (CEPME), France
Ms. Maria Teresa d'Avila
President, AMEP, Portugal
Ms. Anne Löfquist
Product Manager, ALMI, Sweden
Ms. Fotini Legaki
Chairman, HELLAFARM SA, Greece

16:15-16:35 Cyber cafe pause

"Horizon" speaker: **Ms. Linda Tarr-Whelan**
CEO, Center for Policy Alternatives, United States

Discussants:
Mr. Carlo Borgomeo
President, Italian Agency for Youth Entrepreneurship, Italy
Ms. Hella Gerth
Head, Women Entrepreneurs and Spouses, European Commission, Directorate DG XXIII
Mr. Gianpiero Agnello
Corporate Marketing Manager, Credito Italiano, Italy
Mr. Ian Hamilton Fazey
Journalist, Financial Times, United Kingdom

Rapporteurs:
Mr. Erwan Bothorel
General Delegate, EPICEA, France

Ms. Claude Cattan
National Co-ordinator of the NOW Programme, Institute for the Development of Workers' Professional Training (ISFOL), Italy

14.30-18.00

Workshop 6: Access to Markets

⇒ Market Information
⇒ Innovation as a Commercial Opportunity
⇒ Procurement Issues for Women-owned SMEs
⇒ Subcontracting and Business Partnerships

Chair:
Mr. Bernd Schlegel
Director, Deutsche Ausgleichsbank, Germany

Speakers:
Keynote speaker: **Ms. Laura Henderson**
CEO, Prospect Associates, United States

Ms. Nancy Carter
Professor, Marquette University, United States
Mr. Ferran Pedro y Costa
Deputy of Girona, Spain
Ms. Mireille Durand
Manager, Relation Réaction, France
Mr. Antony DeLuca
Director, Office of Small and Disadvantaged Business Unit, US Air Force, United States
Ms. Shigeko Mitsuhashi
President, Tourism Essentials Inc., Japan

16:30-16:50 Cyber cafe pause

"Horizon speaker": **Ms. Judith Maestracci**
Manager, Queensland, Sports Marketing & Management, Australia

Discussants:
Mr. Michael Förschner
Ministry for Labour and Social Affairs, Austria
Ms. Angela Prata Pace
Vice-President, State Chamber of Commerce of Minas Gerais, Brazil
Ms. Vera Yuzkova
BCC Representative, Russian Federation
Ms. Young-Shin Chang
Chairwoman, Korean Businesswomen's Federation, Korea

Rapporteurs:
Ms. Amy Millman
Executive Director, National Women's Business Council, United States

Ms. Maria Gimbert
Consultant, OECD Territorial Development Service/LEED

FRIDAY, 18 APRIL 1997	
MORNING	
9.30- 13.00 11:00- 11:20 Cyber cafe pause	**PLENARY SESSION** *Chair:* **Ms. Anne-Marie Couderc** Deputy Minister for Employment in Charge of Women's Rights, France *Summing Up:* ⇒ Presentation of findings by Workshop Chairs ⇒ Overall conclusions of the conference by the Chair ⇒ Open discussion *Messages of the Conference:* **Ms. Irene Pivetti** Former President of Italian Parliament, Italy **Ms. Kelly Carnes** Deputy Assistant Secretary for Technology Policy, Technology Administration, US Department of Commerce, United States **Ms. Christine Chauvet** General Director, Centre Français du Commerce Extérieur, France *Horizon Speaker:* **Ms. Virginia Littlejohn** President, Global Enterprise Group, United States *Closing Comments:* **Mr. Chris Brooks** Director, OECD Territorial Development Service *Closing Address:* **Ms. Joanna Shelton** Deputy Secretary-General, OECD

CONFERENCE SPONSORS

The OECD would like to express its thanks to IBM for serving as the primary financial supporter of the conference, for creating the Cyber-café for conference participants, and for assisting with the sponsorship of this report. Thanks are also due to a number of other donors, including Westpac Bank (Australia), the Department of Tourism, Small Business and Industry of Queensland (Australia), the Royal Bank of Canada, Credito Italiano (Italy), Havas Voyages (France), Sogeres (France), and the Inter-American Development Bank.

IBM

Westpac

Inter-American Development Bank

ROYAL BANK OF CANADA

Credito Italiano

QUEENSLAND GOVERNMENT

Department of Tourism, Small Business and Industry

HAVAS VOYAGES

AMERICAN EXPRESS

Voyages d'Affaires

SOGERES

OECD PUBLICATIONS, 2, rue André-Pascal, 75775 PARIS CEDEX 16
PRINTED IN FRANCE
(92 98 03 1 P) ISBN 92-64-16040-X – No. 49942 1998